CONTROLLING STEADY-STATE AND DYNAMICAL PROPERTIES OF ATOMIC OPTICAL BISTABILITY

CONTROLLING STEADY–STATE AND DYNAMICAL PROPERTIES OF ATOMIC OPTICAL BISTABILITY

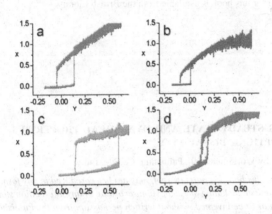

Amitabh Joshi
Eastern Illinois University, USA

Min Xiao
University of Arkansas, USA

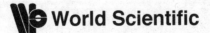 **World Scientific**

NEW JERSEY · LONDON · SINGAPORE · BEIJING · SHANGHAI · HONG KONG · TAIPEI · CHENNAI

Published by

World Scientific Publishing Co. Pte. Ltd.
5 Toh Tuck Link, Singapore 596224
USA office: 27 Warren Street, Suite 401-402, Hackensack, NJ 07601
UK office: 57 Shelton Street, Covent Garden, London WC2H 9HE

British Library Cataloguing-in-Publication Data
A catalogue record for this book is available from the British Library.

**CONTROLLING STEADY-STATE AND DYNAMICAL PROPERTIES
OF ATOMIC OPTICAL BISTABILITY**

ISBN-13 978-981-4307-55-0
ISBN-10 981-4307-55-6

Printed in Singapore by B & Jo Enterprise Pte Ltd

We dedicate this book to our parents

Preface

Atomic optical bistability (AOB) has been a very active field of research in eighties because of its potential application in all-optical logic and switching devices. Bistability was first observed in a passive atomic medium of sodium vapor in 1974 and since then a wide range of activities started in this field of research, and many nonlinear materials including semiconductors were investigated for bistability. The main research was centered on optimizing the size of devices, switching times, operating powers, operating temperatures and other operating conditions. There have been a comprehensive book and some review articles for AOB in two-level systems, such as the excellent book by H. Gibbs: Optical Bistability: Controlling light with light (Academic Press, 1985) and a review article by L. A. Lugiato: Theory of optical bistability in 'Progress in Optics', Vol. 21, p.71 (edited by E. Wolf, North Holland, Amsterdam, 1984). Also, the phenomenon of optical bistability has been discussed in several text books, such as 'Nonlinear Optics' by R. W. Boyd (Elsevier, 3rd Edition, 2008).

In the decade of nineties a great deal of research activities have been seen in the area of multilevel atomic systems exhibiting the phenomenon of electromagnetically induced transparency (EIT). The multilevel atomic system has several advantages over the two-level system as absorption, dispersion, and nonlinear optical properties of one optical field coupled to one atomic transition can be greatly modified by another optical fields coupled to nearby atomic transitions due to the induced atomic coherences. By using such changes in linear and nonlinear optical properties around resonance, due to the EIT, it is easy to manipulate and control nonlinear optical processes in the multilevel atomic systems. Such controls are not available in the two-level atomic system, which have stimulated new research interests in the area of AOB using multilevel atomic systems inside the optical cavity. Hence, one of the new research areas is the controllability of AOB in a three-level atomic system inside an optical cavity, emphasizing the enhanced nonlinear optical process due to induced atomic coherence in such intracavity EIT medium. All-optical switching using AOB and dynamical hysteresis in AOB were studied using the three-level atomic system. Investigations of controlled optical dynamic instability and chaos in the cavity output field with three-level atoms inside an optical cavity were carried out. Phenomenon of stochastic resonance in AOB and noise-induced transitions became easily accessible experiments because of the enhanced nonlinearity in the intracavity EIT medium. Some of those effects have not been observed previously in two-level AOB systems.

The book presents the descriptions of unique controllabilities available in the multilevel EIT systems for studying AOB and related effects, along with the underlying physics encompassing those phenomena. The book is aimed at both graduate level courses and beginning researchers in these areas of AOB and related effects. It can also serve as a reference book for many interesting aspects of nonlinear optical phenomena. We first introduce the background materials for two- and three-level atomic systems useful in understanding optical bistability, multistabilities, dynamic instability, chaos, stochastic resonance, and all-optical as well as noise-induced switching. The book is written at a level comprehensible to readers with the background of a beginning graduate student in the disciplines related to physics and electrical engineering having some good understanding in mathematics, and electricity-magnetism. The other purpose of the book is to develop in detail the central results of optical bistability and associated phenomena with multilevel atomic media inside optical cavity. Through studying this book, the reader should develop a working understanding of fundamental tools and results of this exciting field, and be able to carry out independent research in areas related to AOB and other nonlinear optical processes in multilevel systems.

The basic structure of the book is as follows. Chapter 1 gives an introduction and fundamental concepts of interaction of electromagnetic field with two- and three-level atomic systems, electrical susceptibilities, nonlinearities of these systems, highlighting some open issues along the way. A broad overview for the main concepts and results of multilevel EIT systems, including controllable linear absorption and dispersion properties in three-level atomic systems are presented. The in-depth fundamental notions of AOB along with the historical development for the simple case of two-level atoms are reviewed in Chapter 2. In Chapter 3 the discussion on cavity line-narrowing effects and cavity ring down spectroscopy with three-level EIT systems are given. This Chapter also provides formulation of steady-state AOB with three-level atoms inside an optical cavity, which is followed by descriptions of recent experiments showing controls of shape and rotation direction of the hysteresis cycles by the coupling beam frequency detuning and other experimental parameters. The background material and origin of optical multistability in two- and three-level atoms in terms of the general principle are discussed in Chapter 4. The discussion is further enhanced by incorporating some successful experimental demonstrations of optical multistability in the laboratory.

Chapter 5 describes theoretical and experimental investigations of controlled optical dynamic instability and chaos in the cavity output field with three-level atoms inside an optical ring cavity. The route to chaos via period doubling and its characterization with the help of diagnostic recipes of Fourier transform and Lyapunov exponent calculations are discussed.

In Chapter 6, experiments describing deterministic switching are discussed. The basics of all-optical switching phenomena and their controls using Kerr nonlinearity are described, which were demonstrated in the basic three-level AOB system inside an optical cavity. In such experiments the cavity (probe) output field intensity could be controlled by the coupling field intensity or frequency detuning. The importance of such all-optical switching in optical computation is also highlighted.

The dynamical hysteresis, stochastic resonance and noise-induced transitions in the three-level AOB system are explained in Chapter 7. The formulations of these phenomena, both qualitatively and with simplified mathematical models, are presented in this Chapter. Then, a few novel experiments done in laboratory are discussed.

Some concluding remarks and outlooks are given in Chapter 8 which summarizes this book with emphasis on the unified central theme: controllable cavity output (both steady-state and dynamical behaviors) with a modified linear and nonlinear intracavity medium of an ensemble of three-level EIT atoms.

We would like to thank many colleagues who have worked with us over the years on the topics presented in this book, especially H. Wang, D. J. Goorskey, W. Yang, and A. Brown. We also thank our families for their supports and understanding.

Amitabh Joshi
Min Xiao

Contents

Chapter 1

Introduction

1.1 Background

Atomic optical bistability (AOB) with two-level atoms confined in an optical cavity was extensively studied in the decades expanding eighties and early nineties. The main interest of this study arose due to the potential applications of AOB in all-optical switches, optical transistor-like devices, optical memories, and eventually all-optical information processing circuits [1,2]. The phenomenon of AOB was recognized in two different categories, i.e., absorptive AOB and dispersive (refractive) AOB [3], both of which were studied with a number of theoretical models and also observed in experiments with several atomic and semiconductor solid state systems [1,2]. These two kinds of AOB basically differ from each other due to the different mechanisms involved in their generation. It is the saturation of the atomic transition, which is responsible for the generation of absorptive AOB [4,5]. However, to realize dispersive AOB, the atomic media should have nonlinearity or posses intensity-dependent refractive index [1–8]. Many interesting and ground breaking theoretical works were carried out in two-level AOB systems over the years. The transmission, fluorescent and absorption spectra from optical cavities containing such two-level atomic systems were studied [9–11] using the mean field model and quantum statistical treatment. The two-level atomic AOB system is found to be analogous to first-order phase transition because the diffusion constant of the Fokker-Planck equation, which describes this AOB system quite well, is intensity dependent [5,12,13]. Another important effect predicted in AOB of two-level system was the self-pulsing effect obtained under the steady-state conditions. This effect was obtained through the analytic solutions of the governing equations of such system [14]. A very interesting effect of instability in the upper branch of AOB curve was also discovered [15]. A systematic study had been carried out to determine the domain of instability in the detuning space for a two-level AOB system for the plane wave and Gaussian field profiles [16].

Many novel experiments were also reported in the two-level AOB systems besides the above mentioned theoretical studies. In a sodium atomic beam interacting with the cavity mode field, the single-mode instability was observed. It was possible to directly compare the experimental results so obtained with the theoretical model in all details so that behavior of spontaneous output oscillations generated by such instability was clearly explained [15,16]. One experiment using cold atomic cloud of cesium atoms reported AOB and instability [17]. In that system the basic excitation mechanism of cesium atoms was through their Zeeman sublevels interacting with a circularly-polarized laser beam in an optical cavity. A single laser beam was employed for both optical pumping and saturation. The basic limitation in such experiments on AOB using two-level system is the lack of controllability by external means [5,17]. No methodology was provided in that experiment for locking the optical cavity in observing AOB and instability [17].

Another interesting phenomenon observed in multilevel atomic systems and studied quite extensively over the last two decades is the electromagnetically induced transparency (EIT) due to atomic coherence. The generated atomic coherence in such EIT systems induces changes in the absorption and dispersion properties of the atomic medium and also enhances the nonlinearity in multilevel atomic systems [18-25]. Slowing down of the group velocity of light was also observed in these multilevel EIT systems [25]. The EIT can influence many other physical processes [5], for example, it can cause enhancements of nonlinear processes in harmonic generation [26-28], two-photon absorption [29] and four-wave mixing [30-32]. One of the very interesting and significant experiments in recent years has been the measurement of third-order Kerr nonlinear index of refraction, i.e., the coefficient of the intensity-dependent refractive index in the three-level atomic system [33]. The Kerr nonlinear coefficient was not only measured [5] in the proximity of the EIT condition but also near the more general coherent population trapping (CPT) conditions, for the atomic medium consisting of rubidium atoms in a three-level Λ-type configuration confined in an optical cavity [33,34]. The EIT and CPT conditions lead to enhancement of the third-order Kerr nonlinear index of refraction in such coherent atomic medium. In the three-level atomic system in Λ-type configuration there are two transitions called the probe field transition and coupling field transition. Under the resonant conditions for these two transitions, i.e., when the frequency detunings of both transitions are zero then the coefficient of Kerr nonlinear index of refraction is zero at the exact EIT or CPT condition. However, a very small change in any of these frequency detunings from zero can greatly enhance the value of nonlinear refractive index of this EIT medium in comparison to the two-level atomic medium [5]. How the value of nonlinear refractive index changes with laser intensities and frequency detunings of both probe and coupling laser beams had been systematically studied in the atomic medium consisting of three-level atoms in Λ-type configuration [5,33,34]. During such measurements in those works the bistable behavior of the atomic system was also observed. These experiments clearly demonstrated that AOB in three-level Λ-type EIT system can easily be controlled by frequency detuning and intensity of the coupling field. In fact, the coupling field acts as the controlling field to change the requisite absorption, dispersion and nonlinearities of the atomic medium so that the hysteresis cycle in AOB can be manipulated in its shape and width very easily. Such advantage is not available with the two-level AOB systems [1,2,6,7,15,16] including the systems with multi-Zeeman levels employing only one laser beam [17]. Another important merit of using three-level atomic system in AOB experiments is its simplicity in design. Here in the experiments one can use atoms in a vapor cell [5]. There is no requirement of using atomic beams or cooled atomic sample under vacuum to observe AOB. The three-level Λ-type system can work in a two-photon Doppler-free configuration by sending the probe and coupling lasers beams in the same direction to overcome the first-order Doppler effect [23,24]. The enhancement in the nonlinearity in this three-level EIT system is because of induced atomic coherence among the two-photon connected energy levels. Such atomic coherence can be associated to the effect observed in a previous experiment using Fabry-Perot (FP) cavity filled with sodium atomic vapor [35]. In that experiment the counter-propagating beams provided the photons for two required frequencies of transitions for the atoms in a particular group of velocity distribution [5]. The three-level systems were modeled theoretically over the years and many interesting results were obtained. The phenomenon of spontaneously generated coherence has also found its place in the studies of three-level AOB. It was shown in some earlier and recent studies that the EIT, atomic coherence, quantum interference, and spontaneously generated coherence can effectively control both AOB as well as multistability in three-level systems [36-41].

The necessity to control AOB systems is because of its potential practical applications in in all-optical switches, all-optical memories, optical transistors, and all-optical logic gates and processors. The main advantage of all-optical systems is that there is no necessity for optical-electronic-optical conversion of signal information during the signal processing. All-optical devices use light beams and a nonlinear medium where Kerr nonlinearity in the three-level atomic system can be greatly enhanced due to atomic coherence. Hence, the light intensities required to switch 'on' and 'off' such devices, i.e., the switching thresholds, reduce down appreciably, leading to easy and effective control, which is a crucial requirement for optical signal processing in a faster and reliable manner using very low intensity of light [5]. By controlling the linear and nonlinear optical properties of the three-level atomic medium inside an optical cavity experimentally it is possible to change, modify and control the steady-state and dynamical behaviors of the composite system. For example, by variation of the intensity or beam frequency of the controlling (coupling) laser it is possible to change the threshold values and the shape of the steady-state AOB hysteresis curve [42]. It is also possible to observe transitions from bistable to multistable hysteresis curves using such variations of experimental parameters [43]. An unusual behavior in AOB was also observed, i.e., the rotation of the hysteresis loop changed from usual forward direction (counter-clockwise rotation) to backward direction (clockwise rotation) [44] just by increasing the frequency detuning of the coupling laser beam. Such behavior has been commonly reported earlier in magnetic, electrical, and biological subsystems. The reason of this unusual behavior is from the point of view of energy dissipation considerations. Another very interesting behavior of dynamical hysteresis cycle in the AOB of the three-level EIT atomic system was observed [45]. The dynamical hysteresis cycle is observed when change in input intensity of the cavity field is varied in a non-adiabatic manner. In the composite system of cavity and three-level atomic medium, the output field of cavity shows unstable behavior for certain parametric ranges and exhibits the interesting phenomena of dynamic instability and chaos [46,47]. The main physical reason for the phenomenon of dynamic instability can be attributed to the competition between the two processes occurring simultaneously, i.e., the optical pumping in the coupling transition and the saturation in the probe transition [46]. The phenomenon of instability can be controlled by manipulating either the coupling beam frequency detuning or intensity. The phenomenon of stochastic resonance [48] was also observed in the three-level AOB system as it was easy to control the shape and width of the hysteresis cycle by experimental parameters in such system. Noise-induced switching phenomenon [49] in such AOB systems with three-level atoms could be studied in experiments. This phenomenon occurs due to fluctuations in the atomic coherence caused mainly by laser frequency fluctuations and hence not so easy to realize when two-level atoms are considered as the intracavity medium [5]. The controllability of optical properties of the three-level atom using coupling laser beam is the greatest advantage over the two-level atom.

In this monograph studies of controlled nonlinear optical processes of three-level atoms inside an optical cavity, e.g., AOB and related phenomena, are presented. To be more specific controllable nonlinear optical processes in the multilevel EIT systems contained in an optical cavity for observing AOB, dynamical hysteresis, multistability, dynamic instabilities, chaos, stochastic resonance and noise-induced switching, are the main focus of discussions here. Also, given at places are comparison of them with two-level AOB systems. There are several other interesting nonlinear optical phenomena in related systems which are not discussed because of the limited scope of this monograph.

In this chapter we will first discuss the interaction of a two-level atomic system with a monochromatic coherent field in terms of the density-matrix formulation. The model involves use of the electric dipole and rotating-wave approximation and provides analytic expressions for absorption and dispersive properties of the two-level atomic system. The explicit formulae for the first- and third-order electrical susceptibilities are obtained. The Doppler effect in the two- and three-level systems are discussed. Optical Fabry-Perot cavity and ring cavity are described by deriving their transmission function. Effects caused by intracavity medium on the transmission function are identified. Controls of linear and nonlinear susceptibilities in three-level atomic systems with experimental parameters are also discussed.

1.2   Nonlinearity in a Two-level Atomic System

The interaction of a collection of two-level atoms with a monochromatic coherent light field is an approximation to real atomic systems, but it does provide description of many physical processes in a very simplified manner. For example, the observed phenomena such as optical saturation, optical Stark effect, power broadening, Rabi oscillations, absorption/dispersion and optical nonlinearities could all be modeled very well using the two-level atom approximation of real atomic systems when interacting with coherent light fields. The two-level approximation involves an atomic system to be consist of just two atomic levels connected by a resonant or near resonant optical field. The two-level atomic model is deprived of other features present in real atoms but can be tackled theoretically without invoking perturbation technique [3], providing explanation for several interesting phenomena. Such treatments of interaction between a two-level atomic system and an electromagnetic field have been extensively dealt in the literature [3,50-53].

1.2.1   *Two-level atoms interacting with a monochromatic field: the density-matrix approach*

In the following we present the density-matrix equations for a two-level atomic system interacting with an electromagnetic field including radiative damping mechanism. We adopt the treatment similar to Ref.[3]. Assuming the atomic system being quantum mechanical in nature and hence it is associated with an atomic wave function $\psi(r,t)$, which evolves according to time-dependent Schrödinger equation

$$i\hbar\frac{\partial\psi(r,t)}{\partial t} = \hat{H}\psi(r,t). \tag{1.1}$$

In the above Schrödinger equation operator $\hat{H}$ is called the Hamiltonian, which can be written as a combination of two parts

$$\hat{H} = \hat{H}_0 + \hat{V}(t), \tag{1.2}$$

where $\hat{H}_0$ is the Hamiltonian for a free atom and $\hat{V}(t)$ is the interaction Hamiltonian of the atom and electromagnetic field. The interaction Hamiltonian in the dipole approximation [3] is given by

$$\hat{V}(t) = -\hat{\mu} \cdot \hat{\mathbf{E}}(t). \tag{1.3}$$

Here $\hat{\mu} = -q\hat{\mathbf{r}}(t)$ is the electric-dipole moment operator, q represents the magnitude of electric charge of the electron, and $\hat{E}(t)$ is the electromagnetic field given as $\hat{\mathbf{E}}(t) = (\mathbf{E}(\omega_P)\, e^{-i\omega_P t} + c.c.)$. For the two-level atomic system considered here we denote the energies of the levels as E_i (i=1,2) $(E_2 > E_1)$. The wave function describing the state of two-level atom can be expressed as

$$\psi(\vec{r}, t) = d_1(t)v_1(\vec{r}) + d_2(t)v_2(\vec{r}), \tag{1.4}$$

in which $d_1(t)$ and $d_2(t)$ are complex numbers called probability amplitudes and the functional variables $v_1(\vec{r})$ and $v_2(\vec{r})$ are eigenfunctions of the time-independent Schrödinger equation $\hat{H}v_i(\vec{r}) = E_i v_i(\vec{r})$ obeying the orthonormal property $\int v_i^*(\vec{r})v_j(\vec{r})d^3r = \delta_{ij}$. The time-dependent Schrödinger equation can be recasted in probability amplitudes as

$$i\hbar\frac{\partial d_i(t)}{\partial t} = \sum_i H_{ij}d_j(t), \tag{1.5}$$

and the corresponding density-matrix for the two-level atom is defined by

$$\rho = \begin{pmatrix} \rho_{11} & \cdot\ \rho_{12} \\ \rho_{21} & \rho_{22} \end{pmatrix} = \begin{pmatrix} d_1 d_1^* & d_1 d_2^* \\ d_2 d_1^* & d_2 d_2^* \end{pmatrix}, \tag{1.6}$$

where $\rho_{ji} = \rho_{ij}^*$, and ρ_{ij}^* is the complex conjugate of ρ_{ij}. The expectation value of any observable quantity $\hat{O}$ can be estimated using wave function of the system as

$$< \hat{O} > = \int \psi^* \hat{O}\psi d^3r = \sum_{mn} d_m^* d_n\, O_{mn}. \tag{1.7}$$

Alternatively,

$$< \hat{O} > = \sum_{mn} \rho_{mn}O_{nm} = tr(\hat{\rho}\hat{O}), \tag{1.8}$$

in which $\hat{\rho}$ represents the density operator with components given by $\rho_{mn} = d_n^* d_m$ and $\hat{\rho}\hat{O}$ describes operator product of $\hat{\rho}$ and $\hat{O}$. The time dependence of the density-matrix elements is obtained by the equation $\dot{\rho}_{ij} = d_j^* d_i$ (where dot (.) over the quantities represent their time derivative). By using Eq.(1.5) it is straightforward to obtain [3]

$$\dot{\rho}_{ij} = \frac{i}{\hbar}\sum_k (\rho_{ik}H_{kj} - H_{ik}\rho_{kj}). \tag{1.9}$$

Summing over index k yields

$$\dot{\rho}_{ij} = \frac{i}{\hbar}(\hat{\rho}\hat{H} - \hat{H}\hat{\rho})_{ij} = -\frac{i}{\hbar}[\hat{H}, \hat{\rho}]_{ij}, \tag{1.10}$$

in which the commutator bracket of the operators $\hat{P}$ and $\hat{Q}$ is given by $[\hat{P},\hat{Q}] = \hat{P}\hat{Q}-\hat{Q}\hat{P}$. Equation (1.10) is the evolution equation of the density-matrix in time under the influence of Hamiltonian $\hat{H}$. So far we have not included any decay processes in the density-matrix equations. There are several ways to include such processes in the density-matrix equations. The simplest way is to include them in the phenomenological way. The basic assumptions involved in this method are as follows [3]. The damping of the diagonal elements of the density matrix are related to population decay from the higher energy level to lower energy level. On the other hand the dephasing of off-diagonal elements have no such restriction. By inclusion of the damping effects the density matrix-equation of motion goes as [3]

$$\dot{\rho}_{jj} = -\frac{i}{\hbar}[\hat{H},\hat{\rho}]_{jj} + \sum_{E_i > E_j} \Gamma_{ji}\rho_{ii} - \sum_{E_i < E_j} \Gamma_{ij}\rho_{jj},$$

$$\dot{\rho}_{ji} = -\frac{i}{\hbar}[\hat{H},\hat{\rho}]_{ji} - \gamma_{ji}\rho_{ji}, \quad j \neq i. \tag{1.11}$$

The parameter Γ_{ji} is the measure of population decay rate from level i to j and γ_{ji} is related to the decay rate of the coherence term ρ_{ji}. The term γ_{ji} can be expressed as $\gamma_{ji} = \frac{1}{2}(\Gamma_j + \Gamma_i) + \gamma_{ji}^C$ for an open two-level system, in which parameters Γ_i and Γ_j represent the total population decay rates from levels i and j, respectively and γ_{ij}^C is the dephasing rate due to elastic collisions.

The atomic wave function (Eq.(1.4)) has a definite parity so that diagonal matrix elements of $\hat{\mu}$ vanish due to the symmetry, i.e., $\mu_{11} = \mu_{22} = 0$ and consequently $V_{11} = V_{22} = 0$. However, the off-diagonal elements are non-vanishing and are given [3] by $V_{21} = V_{12}^* = -\mu_{21}\hat{E}(t)$.

The phenomenon of the radiative decay from the upper level (level 2) to the lower level (level 1) is called spontaneous emission. For the closed two-level atomic system the upper level radiatively decays to the lower level at a rate Γ_{21} and hence the radiative lifetime of the upper level is $\tau_1 = 1/\Gamma_{21}$. The dephasing of atomic diploe moment (given by the off-diagonal element ρ_{21}) in the characteristic time τ_2 gives rise to a transition linewidth γ_{21} such that $\gamma_{21} = 1/\tau_2$. Here the dephasing caused by collisions is neglected. For the pure radiative damping case in the two-level system $\tau_2 = 2\tau_1$. The off-diagonal element ρ_{21} term is also governed by the coherence of the two-level system.

In general, damping of a system is described by its interaction with a thermal reservoir with a large number of degrees of freedom. The thermal reservoir may consist of a large number of simple harmonic oscillators or two-level systems (i.e., atomic reservoir). The equations of motion for the system of interest are then usually obtained after tracing over the reservoir variables. There are several ways of dealing with radiative damping. The important ones among others are quantum theory of damping involving density operator and wave function approach, as well as Heisenberg-Langevin approach [51,52,53]. However, these radiative damping mechanisms can be added phenomenologically in the density-matrix equations (as shown above in Eq.(1.11)) of the two-level system to give

$$\dot{\rho}_{22} = -\Gamma_{21}\rho_{22} - i\hbar^{-1}(V_{21}\rho_{12} - \rho_{21}V_{12}),$$
$$\dot{\rho}_{11} = \Gamma_{21}\rho_{22} + i\hbar^{-1}(V_{21}\rho_{12} - \rho_{21}V_{12}),$$
$$\dot{\rho}_{21} = -(i\omega_{21} + \gamma_{21})\rho_{21} + i\hbar^{-1}V_{21}(\rho_{22} - \rho_{11}). \tag{1.12}$$

The system is closed and we can observe the condition $\dot{\rho}_{22} + \dot{\rho}_{11} = 0$. Since $\hat{\rho}$ represents the probability of occupation hence $\rho_{11} + \rho_{22} = 1$ which is known as trace condition on the density operator.

The simplest but very interesting problem is to find the steady-state response of a closed two-level atom in a monochromatic field. We rewrite density-matrix equations (Eq.(1.12)) in a slightly different and general form [3]

$$\frac{d}{dt}(\rho_{22} - \rho_{11}) = -\Gamma_{21}[(\rho_{22} - \rho_{11}) - (\rho_{22} - \rho_{11})^{eq}] - 2i\hbar^{-1}(V_{21}\rho_{12} - \rho_{21}V_{12}),$$
$$\frac{d}{dt}\rho_{21} = -(i\omega_{21} + \gamma_{21})\rho_{21} + i\hbar^{-1}V_{21}(\rho_{22} - \rho_{11}). \tag{1.13}$$

The term $(\rho_{22} - \rho_{11})^{eq}$ is introduced here as the population difference of two levels in thermal equilibrium could be other than -1 [3]. The interaction Hamiltonian in dipole approximation has the following non-zero matrix elements

$$V_{21} = V_{12}^* = -\mu_{21}(Ee^{-i\omega_P t} + E^*e^{i\omega_P t}). \tag{1.14}$$

The exact solution of the density-matrix equations for V_{21} given above is extremely difficult. However, it is possible to solve them under the so called rotating-wave approximation (RWA). In the absence of interaction Hamiltonian ($V_{21} = V_{12}^* = 0$), we find that ρ_{21} has a tendency to evolve in time as $\exp(-i\omega_{21}t)$. So, when ω_P is near to ω_{21} then the portion of V_{21} oscillating as $\exp(-i\omega_P t)$ becomes more important than the other portion going as $\exp(i\omega_P t)$. This means we retain only those terms in the density-matrix equations varying (slowly) as $\exp(\pm i(\omega_P - \omega_{12})t)$ and neglect rapidly varying terms like $\exp(\pm i(\omega_P + \omega_{12})t)$ [3,50,53]. This is called RWA and thus under the RWA, we can write

$$V_{21} = V_{12}^* = -\mu_{21}Ee^{-i\omega_P t}. \qquad (1.15)$$

Applying RWA, the density-matrix equations become

$$\frac{d}{dt}(\rho_{22} - \rho_{11}) = -\Gamma_{21}[(\rho_{22} - \rho_{11}) - (\rho_{22} - \rho_{11})^{eq}]$$

$$+2i\hbar^{-1}(\mu_{21}Ee^{-i\omega_P t}\rho_{12} - \rho_{21}\mu_{12}E^*e^{i\omega_P t}),$$

$$\frac{d}{dt}\rho_{21} = -(i\omega_{21} + \gamma_{21})\rho_{21} - i\hbar^{-1}\mu_{21}Ee^{-i\omega_P t}(\rho_{22} - \rho_{11}).$$

$$(1.16)$$

Clearly, one can see that in RWA ρ_{21} and $\rho_{22} - \rho_{11}$ are driven at their natural frequencies. For example, ρ_{21} is being driven with its resonance frequency ω_{21} and the population inversion $\rho_{22} - \rho_{11}$ almost at the zero frequency [3]. The solution of above equations can be obtained by making use of a slowly-varying quantity $\tilde{\rho}_{21}$ such that $\rho_{21} = \tilde{\rho}_{21}e^{-i\omega_P t}$ [3]

$$\frac{d}{dt}(\rho_{22} - \rho_{11}) = -\Gamma_{21}[(\rho_{22} - \rho_{11}) - (\rho_{22} - \rho_{11})^{eq}]$$

$$+ 2i\hbar^{-1}(\mu_{21}E\tilde{\rho}_{12} - \mu_{12}E^*\tilde{\rho}_{21}),$$

$$\frac{d}{dt}\tilde{\rho}_{21} = [i(\omega_P - \omega_{21}) - \gamma_{21}]\tilde{\rho}_{21} - i\hbar^{-1}\mu_{21}E(\rho_{22} - \rho_{11}). \qquad (1.17)$$

1.2.2   *Absorption and dispersion spectra in steady-state*

In order to obtain the steady-state solution of above equations, the time derivative or the left-hand side of these equations can be set to zero. The solution of the resultant equations (in original variables) gives [3]

$$\rho_{22} - \rho_{11} = \frac{(\rho_{22} - \rho_{11})^{eq}[1 + \Delta_P^2\gamma_{21}^{-2}]}{1 + \Delta_P^2\gamma_{21}^{-2} + 4\hbar^{-2}|\mu_{21}|^2|E|^2(\gamma_{21}\Gamma_{21})^{-1}},$$

$$\rho_{21} = \frac{\mu_{21}Ee^{-i\omega_P t}(\rho_{22} - \rho_{11})}{\hbar(\Delta_P + i\gamma_{21})}. \qquad (1.18)$$

The laser (frequency ω_P) interacts with the atomic transition $|1> \leftrightarrow |2>$ (frequency ω_{21}) with a frequency detuning $\Delta_P = \omega_P - \omega_{21}$. The electromagnetic field induces polarization in the atomic medium. The field-induced polarization $\mathbf{P}(t)$ can be calculated using the off-diagonal elements of the density-matrix operator as

$$\mathbf{P}(t) = N <\hat{\mu}> = N Tr(\hat{\rho}\hat{\mu}) = N(\mu_{12}\rho_{21} + \mu_{21}\rho_{12}), \qquad (1.19)$$

in which N is the number of atoms per unit volume. The electrical susceptibility χ of the atomic medium can be calculated using the relationship
between polarization magnitude P and the electric field E, i.e.,

$$\chi = P/E = \frac{N|\mu_{21}|^2(\rho_{22} - \rho_{11})}{\hbar(\Delta_P + i\gamma_{21})}. \tag{1.20}$$

By making use of Eq.(1.18) for $\rho_{22} - \rho_{11}$ in Eq.(1.20) and then simplifying
the resulting expression yields [3]

$$\chi = \frac{N|\mu_{21}|^2(\rho_{22} - \rho_{11})^{eq}(\Delta_P - i\gamma_{21})\gamma_{21}^{-2}\hbar^{-1}}{1 + \Delta_P^2\gamma_{21}^{-2} + 4\hbar^{-2}|\mu_{21}|^2|E|^2(\gamma_{21}\Gamma_{21})^{-1}}. \tag{1.21}$$

The quantity $2|\mu_{21}||E|/\hbar = \Omega_P$ is called the resonant Rabi frequency, which
depends on transition atomic dipole moment and the field amplitude. We
give more discussion on the Rabi frequency in the beginning of section 1.5.
The susceptibility now reads as

$$\chi = N|\mu_{21}|^2\frac{(\rho_{22} - \rho_{11})^{eq}\gamma_{21}^{-2}\hbar^{-1}(\Delta_P - i\gamma_{21})}{1 + \Delta_P^2\gamma_{21}^{-2} + \Omega_P^2(\gamma_{21}\Gamma_{21})^{-1}}. \tag{1.22}$$

The electrical susceptibility is a complex quantity and can be separated
into its real and imaginary parts, i.e., $\chi = \chi' + i\chi''$ where χ' and χ'' are
given by [3]

$$\chi' = -\frac{\alpha_0 c}{4\pi\omega_{21}}[1 + \Omega_P^2(\gamma_{21}\Gamma_{21})^{-1}]^{-1/2} \times \frac{\Delta_P\gamma_{21}^{-1}[1 + \Omega_P^2(\gamma_{21}\Gamma_{21})^{-1}]^{-1/2}}{1 + \Delta_P^2\gamma_{21}^{-2}[1 + \Omega_P^2(\gamma_{21}\Gamma_{21})^{-1}]^{-1}},$$

$$\chi'' = \frac{\alpha_0 c}{4\pi\omega_{21}}[1 + \Omega_P^2(\gamma_{21}\Gamma_{21})^{-1}]^{-1} \times \frac{1}{1 + \Delta_P^2\gamma_{21}^{-2}[1 + \Omega_P^2(\gamma_{21}\Gamma_{21})^{-1}]^{-1}},$$

$$\tag{1.23}$$

respectively, where we have defined the parameter α_0 as the line-center
absorption coefficient

$$\alpha_0 = -\frac{4\pi\omega_{21}}{c}[N(\rho_{22} - \rho_{11})^{eq}|\mu_{21}|^2(\hbar\gamma_{21})^{-1}]. \tag{1.24}$$

The width of the absorption line (full width at half maximum) is given by

$$\Delta_{HW} = 2\gamma_{21}\sqrt{(1 + \Omega_P^2(\Gamma_{21}\gamma_{21})^{-1})}. \tag{1.25}$$

This linewidth is better known as power broadened linewidth (due to power
broadening) because of the additional factor $(1 + \Omega_P^2(\Gamma_{21}\gamma_{21})^{-1})^{1/2}$ mul-
tiplying with γ_{21}, which depends on the intensity of the optical field Ω_P^2.
The real and imaginary parts of the susceptibility χ are plotted in Fig.1.1
(a,b) using Eq.(1.23).

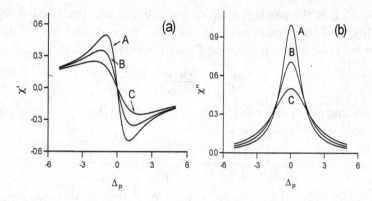

Fig. 1.1 Plot of real (χ') (a) and imaginary (χ'') (b) parts of the susceptibility as a function of atom-laser frequency detuning Δ_P. The curves A, B, and C are for increasing value of $\Omega_P^2(\Gamma_{21}\gamma_{21})^{-1}$.

1.2.3 *First- and third-order susceptibilities and the saturation phenomenon*

It is simple to find that the absorption (χ'') at the line center ($\Delta_P = 0$) decreases by an amount $(1 + \Omega_P^2(\Gamma_{21}\gamma_{21}))^{-1}$, when an intense optical field is used. This phenomenon is known as saturation. We can quantify this behavior by defining the relationship [3]

$$\Omega_P^2(\Gamma_{21}\gamma_{21})^{-1} = |E|^2/|E_{sat}|^2, \tag{1.26}$$

in which the quantity E_{sat} is called the line-center saturation field strength, given by

$$|E_{sat}|^2 = \frac{\hbar^2\Gamma_{21}\gamma_{21}}{4|\mu_{21}|^2}. \tag{1.27}$$

Incorporating the above definition of saturation the expression of susceptibility reads as

$$\chi = -\frac{\alpha_0 c\gamma_{21}^{-1}}{4\pi\omega_{21}} \frac{(\Delta_P - i\gamma_{21})}{1 + \Delta_P^2\gamma_{21}^{-2} + |E|^2/|E_{sat}|^2}. \tag{1.28}$$

Physically the saturation field means that the absorption realized by the optical beam at the line center ($\Delta_P = 0$) of the atomic system reduces to one-half of its usual weak-field value when the optical beam has a field strength of E_{sat}. By expanding the quantity $|E|^2/|E_{sat}|^2$ [3] using a series expansion and retaining only the first and second terms we can rewrite the susceptibility as

$$\chi \simeq -\frac{\alpha_0 c\gamma_{21}^{-1}}{4\pi\omega_{21}} \frac{(\Delta_P - i\gamma_{21})}{1 + \Delta_P^2\gamma_{21}^{-2}} [1 - \frac{1}{1 + \Delta_P^2\gamma_{21}^{-2}} \frac{|E|^2}{|E_{sat}|^2}]. \tag{1.29}$$

We can find the first- and third-order susceptibilities of this system by comparing Eq.(1.29) with the expression $\chi = \chi^{(1)} + 3\chi^{(3)}|E^2|$, which are given by [3]

$$\chi^{(1)} = -\frac{\alpha_0 c \gamma_{21}^{-1}}{4\pi\omega_{21}} \frac{(\Delta_P - i\gamma_{21})}{1 + \Delta_P^2 \gamma_{21}^{-2}} = [\frac{N(\rho_{22} - \rho_{11})^{eq}|\mu_{21}|^2}{\gamma_{21}^2 \hbar}]\frac{(\Delta_P - i\gamma_{21})}{1 + \Delta_P^2 \gamma_{21}^{-2}},$$

$$\chi^{(3)} = \frac{\alpha_0 c \gamma_{21}^{-1}}{12\pi\omega_{21}}[\frac{(\Delta_P - i\gamma_{21})}{(1 + \Delta_P^2 \gamma_{21}^{-2})^2}]\frac{1}{|E_{sat}|^2}$$

$$= -\frac{4N(\rho_{22} - \rho_{11})^{eq}|\mu_{21}|^4}{3\Gamma_{21}\gamma_{21}^3\hbar^3}\frac{(\Delta_P - i\gamma_{21})}{(1 + \Delta_P^2 \gamma_{21}^{-2})^2}. \tag{1.30}$$

Note that the second expression in both $\chi^{(1)}$ and $\chi^{(3)}$ above are obtained by substituting the value of α_0 from Eq.(1.24).

1.3 Doppler Effect in Inhomogeneously-broadened Atomic Systems

A finite linewidth is always associated with any absorption or emission line profile of atomic and molecular systems. The most inherent and fundamental broadening mechanism is the natural linewidth of the absorption/emission lines determined by the radiative damping or spontaneous emission phenomenon [50,54]. Hence the natural linewidth of any absorption or emission line is provided by the radiative damping coefficient Γ of the transition or inverse of the spontaneous life time τ. The natural linewidth comes from Heisenberg's uncertainty principle. When an excited state decays with a natural or spontaneous lifetime of τ to ground state, then the energy uncertainty associated with the transition is $\Delta E \sim \hbar/\tau$ and consequently the width of the transition is $1/\tau = \Gamma$. This result can also be obtained by considering atom as a driven damped classical simple harmonic oscillator, i.e., the Lorentz model of atom [50].

There are several other mechanisms [54-57] for line broadening, causing distribution of resonance frequencies of atomic transitions over a certain frequency range on either side of a central frequency ω_{cen}. These mechanisms are called inhomogeneous linewidth broadening mechanisms of the atomic transitions and have basis in finite temperature and pressure of the atomic samples. One of the important inhomogeneous line broadening mechanisms observable in gaseous media is the Doppler broadening arising due to the Doppler effect, which is related to random thermal movements of atoms in atomic vapor. Suppose the atomic vapor having a thermal velocity distribution, e.g., Maxwell-Boltzmann velocity distribution of its constituent atoms at temperature T, then the linewidth resulting from this finite temperature T is called Doppler broadening. Such a collection of atoms have different resonance transition frequencies as the Doppler shift for these atoms is continuously changing because of their different velocities [54-57]. In the lasers using gases or atomic vapors as active lasing media, Doppler broadening becomes prominent line broadening mechanism for the output laser beam. The same is true when a passive medium is used for absorption/emission spectroscopy [54] at finite temperatures. When high precision spectroscopy is carried out, the Doppler broadening mechanism can cause severe restrictions on such works. Several methods have been proposed to circumvent the Doppler broadening mechanism, e.g., by reducing the temperature of gaseous samples or employing laser cooling techniques. Another simple technique to circumvent Doppler broadening is called saturation absorption spectroscopy [54]. Also, when more laser beams interact with multilevel atomic systems, two-photon absorption with specially-arranged beam propagation directions can be utilized to cancel the Doppler effect. The discussion of eliminating first-order Doppler effect in three-level atomic vapor will be discussed subsequently [42].

1.3.1   *Doppler effect in a two-level atomic system*

The Doppler broadening mechanism can easily be observed in dilute atomic vapors in thermal equilibrium. Since atoms are in random thermal motion having a velocity distribution, the frequencies of absorbed or emitted radiations by atoms depend on the individual atomic velocities. Consider an atom (which is assumed to be at rest with respect to the frame of reference of laser) with transition frequency ω_A, which interacts with a laser or a monochromatic electromagnetic field having frequency ω_L. If the atom moves with velocity v_x along x-direction with $v_x << c$, the observed frequency of the electromagnetic field, as seen from the atom, is $\omega'_L = \omega_L(1 - v_x/c)$. The value of ω'_L is higher or lower than ω_L depends on the direction of relative movement of the atom and electromagnetic wave. ω'_L is higher (lower) than ω_L if the atom is moving in opposite (same) direction to that of the electromagnetic wave meaning $v_x < 0$ ($v_x > 0$). The resonance condition of absorption for that particular atom is when $\omega'_L = \omega_A$. This means that we can rewrite $\omega_L = \omega_A/(1 - v_x/c) \simeq \omega_A(1 + v_x/c)$. In other words the resonance frequency of an atom moving with velocity v_x

is $\omega_B = \omega_A(1 + v_x/c)$. Qualitatively, when the atom moves towards (away from) laser, it absorbs blue-shifted (red-shifted) radiation from the laser [54–57].

From above discussion it is clear that atoms respond differently to laser absorption depending on both magnitude and sign of v_x, hence this mechanism belongs to the inhomogeneous line broadening mechanism. The distribution of resonance frequencies in an atomic gaseous sample in thermal equilibrium at temperature T can be calculated by following consideration. The fraction of atoms having a velocity component between v_x and $v_x + dv_x$ in the direction of propagation of the electromagnetic wave can be expressed [54-57] using the Maxwell-Boltzmann velocity distribution function for thermal velocities at temperature T

$$S_v(v_x) = \sqrt{\frac{M}{2\pi k_B T}} e^{-\frac{M v_x^2}{2 k_B T}}. \tag{1.31}$$

In the above expression, the parameter M specifies atomic mass and k_B is Boltzmann constant. The Maxwell-Boltzmann distribution function is Gaussian in nature. Hence, when Doppler broadening is the prominent line broadening mechanism of the system, it give rises to a Gaussian distribution in the spectrum [54-57]. The function $S_v(v_x)$ has been written in its normalized form, i.e., $\int_{-\infty}^{\infty} S_v(v_x) dv_x = 1$. From $\omega_B = \omega_A(1 + v_x/c)$ it is straightforward to get $v_x = (c/\omega_A)(\omega_B - \omega_A)$. The distribution function for ω_B is $K(\omega_B) = S(v_x)|\frac{dv_x}{d\omega_B}|$, which is given by

$$K(\omega_B) = \omega_A^{-1} \sqrt{\frac{Mc^2}{2\pi k_B T}} \exp[-(\frac{\omega_B - \omega_A}{\omega_A})^2 \frac{Mc^2}{2k_B T}]. \tag{1.32}$$

The function $K(\omega_B)$ is clearly a Gaussian distribution function of frequencies. The maximum of this distribution function is

$$\omega_A^{-1} \sqrt{\frac{Mc^2}{2\pi k_B T}}, \tag{1.33}$$

which occurs at $\omega_B = \omega_A$. The full-width-at-half-maximum (FWHM) of this function is given by

$$\Delta_D = 2\omega_A \sqrt{\frac{2 k_B T ln2}{Mc^2}}. \tag{1.34}$$

This width defined through Δ_D is called the Doppler width of atomic transition [54-57]. A typical Doppler limited absorption profile is given in Fig.1.2.

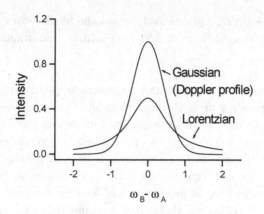

Fig. 1.2   Plot of Doppler-limited absorption profile as a function of $\omega_B - \omega_A$. For the sake of comparison the natural-linewidth-limited profile is also given.

1.3.2   *Doppler effect in three-level atomic systems*

The three-level systems, as shown in Fig.1.3(a,b) interact with two different laser fields at different (or same) frequencies. When a strong coupling laser beam (sometimes also called as a control or pump beam) in the terminology of EIT in the three-level atomic systems, interacts with the atomic transition from state $|3\rangle$ to state $|2\rangle$ in a Λ-type configuration of Fig.1.3(a), or from state $|2\rangle$ to state $|3\rangle$ in a ladder-type configuration of Fig.1.3(b), the absorption and dispersion properties of the probe beam interacting with the transition from state $|1\rangle$ to state $|2\rangle$ will be greatly altered. One of the main observed effects is absorption reduction in the probe beam, which is due to atomic coherence established between states $|1\rangle$ and $|3\rangle$ of atomic systems as a result of interacting with two laser beams. The presence of two laser beams interacting with two different transitions causes destructive interference in the amplitudes governing the two transition probabilities and hence there is reduced absorption and altered dispersive properties. The atomic vapor sample at room temperature possesses Doppler-broadened profile, hence in order to observe reduced probe absorption, a very strong coupling laser beam is required. This was confirmed in the early experiments on EIT using vapors of strontium and lead with strong pulsed lasers [21,22]. It has been shown both theoretically and experimentally that EIT can be obtained with weak CW diode lasers in the three-level atomic systems of Fig.(1.3) provided two-photon Doppler-free configurations are used for such atomic systems [23-25]. The two-photon Doppler-free configuration for the atom-photon interaction can be achieved by allowing the coupling

and probe laser beams to propagate in opposite directions in the atomic
vapor medium made up of three-level ladder-type system [Fig.1.3(b)]. For
the Λ-type three-level system [Fig.1.3(a)] the two laser beams should propagate in the same direction in the atomic vapor medium in order to achieve
Doppler-free configuration [23-25]. Detailed calculations for the modified

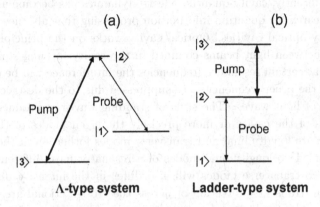

Fig. 1.3   Two different arrangements of three atomic levels known as (a) Λ-type system
and (b) ladder-type system.

absorption and linear index of refraction for the probe laser beam including
Doppler effect are given in Ref.[24]. Simplified closed-form results were obtained in that reference. The effect of residual (second-order) Doppler effect
on EIT and the effect of laser linewidths were also discussed. The experimentally measured EIT results, absorption reduction as a function of probe
beam frequency detuning, with coupling frequency detuning as a parameter were compared with theoretical predictions and excellent agreements
were obtained between them. This kind of quantitative comparisons EIT
and other related measurements made significant impacts in understanding
various effects in EIT systems.

1.4   Optical Cavity

An optical cavity, sometimes also referred to as an optical resonator, is a
device containing mirrors in certain configuration so that it acts either as
a standing wave cavity (or resonator) or ring cavity for light waves [58,59].
Optical resonators are primary components in fabrication of lasers, optical
bistability devices, spectroscopy setups, and cavity-quantum electrodynam-

ics systems etc. Usually, the gain medium of a laser is kept inside the optical cavity, in which light travels back and forth in the cavity and thus furnishing the feedback required to build up laser action [59,60]. Optical cavities are also used in all-optical switching devices, cavity ring-down spectroscopy systems, optical parametric oscillator systems, and interferometric systems etc. Recently, cavity-quantum electrodynamics has become an active area of research for quantum information processing that also needs use of fine quality optical cavities. Optical cavity works on the principle of interference between light beams confined in the cavity forming standing waves. Only at certain resonance frequencies the interference will be constructive, while the other frequencies are suppressed due to the destructive interferences of light waves. The style of standing waves so produced are called modes of the cavity or more precisely the eigen modes of the resonator. There are longitudinal and transverse modes formation in the optical resonator. Two longitudinal modes of a resonator have different frequencies but two transverse modes will also differ in the intensity distribution of light in transverse direction or across the cross-sectional area of the light beam. The fundamental transverse mode of a resonator is called TEM_{00} mode, which has a Gaussian spatial distribution of intensity [58,60] across its cross-section.

Several kinds of optical resonators can be constructed by selecting different mirror arrangements, using different focal lengths and separations of the mirrors. The stability of optical resonators depends crucially on these geometrical parameters. For certain geometrical design of optical cavity, the beam remains sustained in the resonator despite a very large number of feedback from the multiple reflections. Such resonators are stable. However, unstable resonators are also useful in certain applications. Optical resonators are typically characterized by two parameters, e.g., beam waist size and the radius of curvature for the sustained field mode inside the cavity [60].

1.4.1 *Optical Fabry-Perot cavity: transmission and reflection functions, finesse and quality factor*

Optical cavities can be constructed using two plane mirrors or spherical mirrors facing each other and separated by some finite distance. The simplest design of an optical cavity is a FabryPerot (FP) cavity, made of two plane partial reflecting mirrors. This is a typical standing wave resonator.

It is easy to find out the total transmitted and reflected intensities from such FP cavity when a monochromatic plane wave light is injected into this cavity. In Fig.1.4, the physical description of such situation is drawn. A light beam having intensity I_0 with wavelength λ and wave vector $k = \frac{2\pi}{\lambda}$ falls onto a plate as the input beam. The thickness and refractive index of the plate are d and n, respectively. The light (intensity) transmission and reflection from the two surfaces are defined by transmittance (T) and reflectance (R), respectively. The plate material is assumed not to absorb any light, implying R + T = 1. The total transmitted and reflected intensities I_T and I_R from the plate can be calculated by summing up the amplitudes of the respective waves, including appropriately the phase shifts introduced by the plate after each internal reflection. We introduce new parameters r, r', t, and t', where r (t) is the reflectance (transmittance) for the wave amplitude for a wave traveling [58,61] from the surrounding medium (n') to the plate (n) and r' (t') represents the reflectance (transmittance) for a wave going from plate (n) to surrounding medium (n') (see Fig.1.5). Now it becomes easy to calculate the total sums of the reflected and transmitted amplitudes after multiple passages for light beam through the plate. Under the condition of no loss for the beams due to absorption, there is time reversal symmetry in the wave propagation, which makes it quite easy to find relationship between the reflectance and transmittance values for field amplitudes. The pictorial depictions of propagations of such normal and reversed beams through a surface [58,61] are given in Fig.1.6. By comparing the two situations in Fig.1.6, it is easy to get $tt' + rr = 1$ and $tr' + rt = 0$ and hence $r' = -r$, $r'^2 = r^2 = R$, and $tt' = T = 1 - R$.

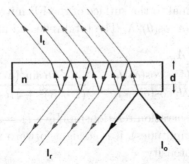

Fig. 1.4 Multiple reflection and transmission from a plane-parallel plate of thickness d leading to interference (adopted from [61] with permission).

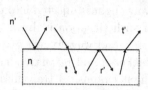

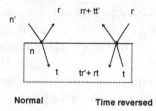

Normal   **Time reversed**

Fig. 1.5 Reflectance and transmittance at an interface between two materials. The material does not absorb the light (adopted from [61] with permission).

Fig. 1.6 Sketch of the input normal and time reversed reflection and transmission at an interface of two materials not absorbing the light (adopted from [61] with permission).

In Fig.1.4, the incident light is oblique, however, in the following the calculations are given for the normal incident light, which is the case for real experimental situations. The sum of all the reflected waves from plate interface provides total reflected amplitude [61]. The waves defined by a tag 'p' at interface pick up different phases due to the different optical path difference $2pdn$. The amplitude of 'p' wave is given by $tt'(r')^{2p-1}$ due to multiple reflections and transmissions. The first reflected wave needs a slightly different treatment while summing up all amplitudes. The sum of all wave amplitudes at the interference gives the expression of total reflected amplitude [58, 61]

$$A_R = A_0 r + A_0 \exp(2ikdn)tt'r' \sum_{p=0}^{\infty}(r')^{2p}\exp(2ikpdn) = A_0 r\frac{1 - e^{i\delta}}{1 - Re^{i\delta}},$$

$$(1.35)$$

in which A_0 is the amplitude of incident wave, k is the wave vector defined as $k = \frac{2\pi}{\lambda}$ and $\delta = 2kdn = 4\pi dn/\lambda$. For the oblique incidence (i.e., $\theta \neq 0$ with reference to normal to the surface of the FP mirror) the expression of δ modifies to $\delta = 4\pi dn\cos(\theta)/\lambda$. The total reflected intensity is given by

$$I_R = A_R \cdot A_R^*$$
$$= \frac{2R(1 - \cos(\delta))I_0}{1 - 2R\cos(\delta) + R^2} = \frac{4R\sin^2(\delta/2)I_0}{(1 - R)^2 + 4R\sin^2(\delta/2)}. \quad (1.36)$$

Using the energy conservation relationship $I_R + I_T = I_0$ (or summing over all the transmitted amplitudes), it is straight forward to get the expression for the transmitted intensity

$$I_T = \frac{T^2 I_0}{1 - 2R\cos(\delta) + R^2} = \frac{T^2 I_0}{(1 - R)^2 + 4R\sin^2(\delta/2)}. \quad (1.37)$$

The above expressions for total reflected and transmitted intensities are related to Airy functions. A quantity called quality factor Q for the FP cavity in terms of its reflectivity (assuming the two interfaces have the same reflectivities) is defined as [61]

$$Q = \frac{4R}{(1-R)^2},$$

(1.38)

leading to

$$I_R = \frac{Q\sin^2(\delta/2)I_0}{1 + Q\sin^2(\delta/2)},$$

$$I_T = \frac{I_0}{1 + Q\sin^2(\delta/2)}.$$

(1.39)

The quality factor Q is related to a Finesse parameter of the FP cavity and it quantifies the energy storage capability of the optical cavity. Usually Q is used to characterize electronic circuits and microwave-related devices. The definition of Q for the optical cavity is $Q = \omega/\delta\omega$. When Q is high then it means high energy storage capability of the cavity (having resonance frequency ω) and at the same time narrow cavity resonance profile (i.e., narrow linewidth ($\delta\omega$) of the cavity transmission profile). In comparison to electrical/electronic circuits, the quality factor of an optical or microwave cavity is usually not so large because the intrinsic dissipation is large under normal conditions. However, the cavity quantum electrodynamics systems in optical physics really needs very high Q cavities for certain kinds of experiments, which have pushed the Q value in optical regime very high. Despite high reflectance possessed by the surfaces of FP cavity, all the incident optical energy can pass through at the transmission peaks. The transmission function (Eq.(1.39)) is plotted in Fig.1.7 for different values of parameter Q (defined in terms of R in Eq.(1.41) in the following) [61]. The x-axis of this plot represents different interference orders $N^i = \frac{\delta}{2\pi}$. It is quite clear from the plots that the width of transmission profile decreases with increased quality factor Q or the transmittance becomes sharper with high Q value of the cavity. Alternatively, with high Q cavities, there is more suppression of light between the two transmission peaks.

The sharpness of the cavity transmission profile can be described by its width, which is related to Finesse of the cavity. The Finesse of an optical cavity is defined as the ratio of peak-to-peak distance to half-intensity width or halfwidth of the peak. The halfwidth of the transmission peak can be obtained when transmittance becomes half of its maximum (or peak) value, i.e.,

$$\frac{1}{1 + Q\sin^2(\delta/2)} = \frac{1}{2}. \tag{1.40}$$

This means $Q\sin^2(\delta/2) = 1$ and for $|\delta - 2N^i\pi| \ll 1$, $\delta = \pm\frac{2}{\sqrt{Q}}$. Since the phase difference between two transmission peaks is 2π, so the Finesse F can be expressed by [58,61]

$$F = \frac{\pi}{2}\sqrt{Q} = \frac{\pi\sqrt{R}}{1 - R}. \tag{1.41}$$

The broadening superimposed on any spectral line by the FP resonator is critically responsible for the resolution of spectral lines provided by the FP resonator. Hence it is important to determine what spectral linewidth $\Delta\lambda$ is equivalent to this instrumental linewidth [61]. Considering the case when light is incident at the normal to the surface, then at the peak of transmission profile, the phase difference δ is given by [61]

$$\delta = 2kdn = \frac{4\pi dn}{\lambda} = 2N^i\pi. \tag{1.42}$$

In Eq.(1.42) above, the parameter N^i characterizes interference order. The instrumental linewidth (in terms of the phase difference), imposed by the FP cavity because of its finite finesse F, is

$$\Delta\delta_{FP} = \frac{2\pi}{F}. \tag{1.43}$$

On the other hand the spectral linewidth, e.g., the capability of distinguishing nearby wavelengths λ and $\lambda + \Delta\lambda$ by the FP cavity, is [61]

$$\Delta\delta = \frac{4\pi dn}{\lambda} - \frac{4\pi dn}{\lambda + \Delta\lambda} \simeq 4\pi\frac{\Delta\lambda dn}{\lambda^2}. \tag{1.44}$$

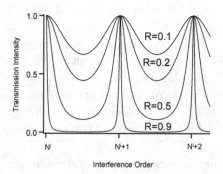

Fig. 1.7   The transmitted intensity of a plane-parallel plate Fabry-Perot cavity for different values of R (or Q) as marked on the transmission profiles. Here large R means large Q.

By equating $\Delta\delta_{FP} = \Delta\delta$, the resolving power of the FP resonator is

$$\frac{\lambda}{\Delta\lambda} = \frac{2Fdn}{\lambda} = FN^i. \tag{1.45}$$

which is a product of Finesse F and the interference order N^i [58,61].

Another important parameter for the FP cavity (or an optical cavity) is called free spectral range (FSR). It is defined as the frequency range of FP cavity in which no overlapping of adjacent orders of transmission peaks. To be specific, the FSR of the FP cavity is the frequency spacing between two successive transmission peaks of orders N^i and $N^i + 1$, and is given by: FSR= $c/(2nd)$. Since $2nd$ is the round-trip path length so it can be easily remembered that the FSR is simply the speed of light divided by the round-trip path length. Alternatively, it is the inverse of the light's round-trip travel time through the FP cavity. It is calculated by equating the phase shifts of two wavelengths (separated by one FSR with each other) λ and $\lambda + \Delta\lambda_{FSR}$, falling at the transmission peaks of interference order $N^i + 1$ and N^i, respectively [58,61]

$$\delta(\lambda) = \delta(\lambda + \Delta\lambda_{FSR}) + 2\pi,$$
$$\frac{2dn}{\lambda} = \frac{2dn}{\lambda + \Delta\lambda_{FSR}} + 1, \tag{1.46}$$

that gives $\Delta\lambda_{FSR} = \frac{\lambda^2}{2dn}$ and that gives $\frac{\Delta\lambda_{FSR}}{\Delta\lambda} = F$.

The transmittance of a FP cavity shows high transmission when the condition for constructive interference at the wavelength of the light is satisfied. No (or very little) light is transmitted under the condition of destructive interference of light. Hence, the FP cavities are useful tool to measure light wavelengths as well as control their values when carrying out some precision optical experiments such as the high resolution spectroscopy. Alternatively, for the known wavelengths of the incident light it is possible to characterize the thickness of the FP cavity or the refractive index of the plate material. In this way the physical parameters pertaining to such cavities can be obtained. Such FP cavity is also known as an optical etalon when used in a laser device to select a single longitudinal mode of the laser oscillation. The etalon also works as wavelength tuning element for laser when kept inside the laser cavity and allowing it to tilt with respect to the direction of propagation of laser beam [61]. Some times air-spaced FP interferometers are also used, where air-pressure in between the end plates are used for frequency tuning in high resolution spectroscopy purposes.

1.4.2 *Optical ring cavity*

The FP resonator discussed above is called standing-wave resonator because
the light beam travels back and forth between two reflectors (mirrors) and
forms standing-wave cavity modes. There is another kind of resonator
used in optical experiments, which is called traveling-wave resonator, or
optical ring resonator. To realize a ring resonator conveniently, typically
three mirrors or reflecting surfaces are required, though, sometimes two
mirrors can also serve the purpose. In the experiments done with three-
level atoms, which will be discussed subsequently, we have always used three
mirror ring resonator. The four-mirror ring resonator is usually discussed
in standard theoretical models of AOB [1,2,4,5]. In such ring resonator the
light wave travels only in a single direction. The reflected and transmitted
intensities from a ring resonator or an optical ring cavity for a plane wave of
monochromatic light have exactly the same expressions as discussed above
for the FP resonator and are given by Eqs.(1.36) and (1.37), respectively.

1.4.3 *Optical ring cavity with an intracavity medium: mod-ification of transmission function*

When an atomic medium is placed inside an optical ring cavity, as shown
in Fig.1.8, the absorption and dispersion properties of the intracavity
medium greatly influence the transmission properties of the cavity given
in Eq.(1.37). The most obvious effect is the suppression of cavity transmission when the atomic medium is highly absorptive at a cavity resonant frequency. The presence of dispersion, i.e., $dn/d\omega \neq 0$, in the atomic medium
causes nearby cavity resonant frequencies to be shifted. The interesting
phenomena arising out of such composite system of atomic medium inside
the optical cavity include optical bistability and related effects which will be
discussed in following chapters. Normally, to observe optical bistability one
needs feedbacks of optical radiation to the intracavity nonlinear medium,
which can be provided by the optical cavity. Since the analysis of FP resonator and the ring resonator is basically identical so in the following we
describe only how the transmission profile of a FP resonator changes with
an intracavity medium. The total transmitted amplitude is calculated by
summing over all amplitudes transmitted out after multiple reflections of
waves inside the resonator. Different waves designated by the parameter
'p' differ in phase due to the different optical path difference $2\,p\,d\,n$ and in
amplitude due to additional reflection and transmission quantified by the

factor $tt'e^{-\alpha L}(r')^{2p}e^{-2p\alpha L}$. The final total transmitted amplitude is thus given by

$$A_T = A_0 tt' e^{-\alpha L} \sum_{p=0}^{p=\infty} (r'e^{-\alpha L})^{2p} e^{ip.2kdn}$$

$$= A_0 T e^{-\alpha L} \sum_{p=0}^{\infty} (Re^{-2\alpha L})^p e^{ip\delta}$$

$$= \frac{A_0 T e^{-\alpha L}}{1 - Re^{-2\alpha L}e^{i\delta}}. \tag{1.47}$$

The expression [corresponding to Eq.(1.37)] for the transmitted intensity thus becomes [62]

$$I_T = \frac{T^2 e^{-2\alpha L} I_0}{(1 - Re^{-2\alpha L})^2 + 4Re^{-2\alpha L}\sin^2(\delta/2)}. \tag{1.48}$$

In above expression the coefficient α represents extinction of light per unit length and the total absorption per round-trip is given by $e^{-2\alpha L}$ for the FP cavity and by $e^{-\alpha L}$ for the ring cavity. So an appropriate factor for absorption per round-trip should be used in Eq.(1.48) depending on the type of cavity. The round-trip phase δ is given more explicitly by

$$\delta = \frac{\omega l}{c} + \frac{(n_0 - 1)\omega L}{c} + \frac{(n_2 I_{cav})\omega L}{c} - 2\pi m, \tag{1.49}$$

where l is the total length of the cavity; L is length of the atomic medium placed inside the cavity, usually in a glass cell; m is an integer. I_{cav} is the intensity of the field inside the cavity, and refractive index $n = n_0 + n_2 I_{cav}$,

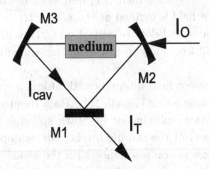

Fig. 1.8 Typical three-mirror optical ring cavity with an intracavity medium. The input mirror M2 and the output mirror M1 have transmissivities T2 and T1, respectively, of the order of 1-2%. The third mirror is highly reflecting and is mounted on a PZT driver for cavity length scanning.

composed of linear part (n_0) and Kerr nonlinear part (n_2) multiplied by the field intensity (I_{cav}) inside the cavity. Eq.(1.48) above is the response function of the three-mirror optical ring cavity with an atomic medium of length L placed in one arm of the cavity. The expressions given in Eqs. (1.48) and (1.49) will be discussed in Chapter 3, on cavity induced line narrowing effect.

1.5   Controllable Linear and Nonlinear Susceptibilities in Three-level Atomic Systems

We have discussed interaction between a two-level system and a coherent radiation field (in subsection 1.2.2), and defined the Rabi frequency $\Omega_P = 2|\mu_{21}||E|/\hbar$. It is the coherent excitation of atom by a monochromatic or near-monochromatic electromagnetic field resonant with an atomic transition that produces such Rabi oscillation. The Rabi oscillation can be observed only when the natural life-time τ of the excited state is much larger compared to inverse of the Rabi oscillation frequency i.e., $\tau > \Omega_P^{-1}$, where $\tau = \Gamma_{21}^{-1}$. The natural or spontaneous decay of the excited state is an incoherent process governed by the radiative decay constant Γ_{21}. Alternatively, when the coherent excitation field is strong, there is a sinusoidal oscillation of the population between the two transition levels with this Rabi frequency. In the discipline of quantum optics, the Rabi frequency is an important quantity and it provides the strength of interaction between coherent field and the atom for a particular transition. The generalized Rabi frequency for a two-level atomic system including frequency detuning of atomic transition with coherent field is defined as $\Omega_{Rabi} = \sqrt{\Omega_P^2 + (\Delta_P/2)^2}$, where Ω_P is as mentioned above and defined after Eq.(1.21), and Δ_P is the detuning of the coherent field frequency from the atomic transition frequency [4,50].

When a strong coherent field interacts with atoms it causes dynamic Stark splitting of the atomic levels [50,53]. The Rabi frequency provides a measure of dynamic Stark splitting or ac-Stark splitting (also known as the Autler-Townes effect) of the atomic levels. For example, when the coherent probe field is exactly on resonance with the atomic transition then the dynamic Stark splitting due to the field is given by the Rabi frequency, in the order of magnitude, provided there is no Doppler broadening. The splitting is symmetric around the resonant transition frequency. In the phenomenon of EIT, Autler-Townes splitting caused by the weak field also plays a crucial role. The quantum interference between the Autler-Townes components enhances the EIT dip. The medium becomes transparent due to the cooperativity of these two effects. The two phenomena, i.e., EIT and Autler-Townes splitting can be distinguished from each other by the level of dephasing occurring in the system. For example, it is the Autler-Townes splitting observed when there is a large dephasing on the unlinked transition but the observation of EIT is difficult.

1.5.1 *Early works on coherent population trapping*

The basic physical mechanism of CPT and EIT (which came into existence much later) could be well understood with the key results of Fano [63], which were obtained in early sixties. Fano used two different excitation paths from the lower state. The first path was from lower state to the ionization continuum state, while the second one was from lower state to an autoionizing state. When atom is excited to the autoionizing state it spontaneously decays to the ionizing state continuum. In this way the final ionizing state continuum can be reached through two different paths and interference between these two transition paths produces asymmetric peaks in the excitation spectrum [4]. The analysis of Fano shows that (i) the transition probability vanishes on one side of the resonance and (ii) there is zero absorption in the medium due to the interference between the transition amplitudes of these two paths. The population trapping in specific levels is the mechanism responsible for zero absorption in the medium [4,63]. The results of Fano gave motivation for the formulation of CPT [18,64] phenomenon. One of the simplest system in which CPT can be observed is a three-level system interacting with two coherent fields on its two transitions. The first experimental observation of CPT was made in sodium atomic vapors using its three-levels in Λ-type configuration [64]. In this system the fluorescence disappeared as the population got trapped in two ground states due to Fano interference of transition probability amplitudes and hence no excitation to the excited state was possible. It can also be understood by assuming atom initially in the superposition of its two ground states and interacting with a coherent field such that the probability amplitude for being in the upper state becomes zero, so the population does not move out from two lower ground states [4]. The physical basis of CPT involves the destructive interference between the two paths connecting the excited state $|e\rangle$ to two ground states $|g\rangle$ and $|g'\rangle$. It is also possible to provide an explanation for CPT in terms of two eigenstates for the Hamiltonian describing the atom-field system and their interaction. These two eigenstates named as 'bright state' and 'dark state', are two different coherent superpositions of the ground states. There is an electric-dipole allowed coupling between upper excited state and the bright state but the dark state is not connected to the upper excited state by any interaction. This kind of situation is obtained when two laser fields interacting with three-level system are in just right ratio of their field amplitudes so that it causes the expectation value of dipole moment from the dark state to the excited state $|e\rangle$ to be zero, i.e., $\langle Darkstate|\mu'|e\rangle = 0$ and μ' is the dipole moment of the transition [4]. The dark state $|Darkstate\rangle$ can be formed, in ideal case with no decay, at any ratio of two Rabi frequencies. When these two transition Rabi frequencies are of the same order, the CPT conditions are prevailing. On the other hand, if one is much bigger than another, it is EIT. The CPT has been extensively studied both in the theory as well as in the experiments [18,64] over the years.

1.5.2   *Coherent population trapping vs electromagnetically induced transparency*

The CPT phenomenon is quite general but EIT is a small portion of that phenomena, for certain parametric range. The magnitudes of the two coherent fields (or two Rabi frequencies) employed in original CPT experiments were of the same order and thus the interference effects arose from both fields. Usually when systems display EIT, one of the fields, usually called probe field, is much weaker than the another field called the coupling field, and one only concerns how the absorption and dispersion properties of the weak probe field are influenced by the strong coupling field. Usually in CPT studies, the lower two levels participating in the three-level Λ-type system are either Zeeman or hyperfine levels of the ground states of the atom and both have some population. On the contrary, in EIT studies one of the lower levels possesses basically negligible population due to the strong optical pumping provided by the coupling field. The EIT can be explained using a dressed-state analysis of the system. Dressed-states mean that the original (bare) atomic states are dressed with an electromagnetic field, in other words they are eigenstates of the time-independent total Hamiltonian, including atom-field interactions [65]. The Hamiltonian is composed of two parts, i.e., the bare atomic Hamiltonian as well as the atom-field interaction. In EIT configuration it is the coupling field, which causes formation of two dressed states for the excited state. When there is a destructive interference between the two probe transition amplitudes from the lower state to upper two dressed states, the phenomenon of EIT occurs [4].

The EIT phenomenon can also be explained by descriptions other than the dressed-state picture. The phenomenon of quantum interference can produce coherence in the atom-field system and EIT can be explained by using such induced coherence. Note that the coherence in the coupled system can be estimated using off-diagonal elements of the density operator of the system. In semiclassical picture of atom-field interaction, coherence is directly related to the oscillating electric dipole due to the interaction of coupling field with a pair of quantum states. If the oscillating dipole can be excited in several different ways, an interference arises between several such contributions to this dipole. The total electric dipole then can be found by summing over all these contributions and that provides the explanation of EIT similar to what is done to understand Fano interference in autoionization [4,63].

1.5.3 *Controlling linear absorption and dispersion properties in three-level electromagnetically induced transparency systems*

In the above discussions some qualitative differences between EIT and CPT are brought out. The simplest systems to observe both EIT and CPT are the three-level systems, which can be found in three different configurations

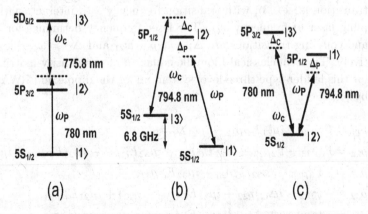

Fig. 1.9 Typical three-level Rb atomic systems in (a) ladder-type configuration; (b) Λ-configuration; and (c) V-type configuration.

known as cascade or ladder-type configuration, Λ-type configuration, and
V-type configuration. In the following, we will concentrate only on the EIT
situation, since the experiments described in this monograph all fall into
this category. These three basic configurations are shown in Fig.1.9. In
these three-level systems, two pairs of levels ($|1\rangle \leftrightarrow |2\rangle$) and ($|2\rangle \leftrightarrow |3\rangle$)
are connected by dipole allowed transitions. For the third pair of levels
($|1\rangle \leftrightarrow |3\rangle$), the dipole induced transition is forbidden. EIT phenomena
have been extensively explored and studied both in ladder and Λ-type configurations of three-level systems. There is no need to do additional optical
pumping when such three-level configurations of three-levels are used for
real atomic systems (due to the commonly existed ground-state fine structures in alkali atoms). To describe the phenomenon of EIT quantitatively
for both ladder and Λ-type configurations, the density-matrix approach is
quite convenient as this approach can incorporate radiative decays or other
decay mechanisms responsible for population decays and decoherences in
a very natural way. There is another approach to study such EIT using
probability amplitudes for the energy states. Both of these approaches give
identical final results [66-68].

A. Three-level system in ladder configuration

The first three-level system we consider is in ladder-type configuration,
where the energy levels are given by E_i (i=1,2,3) such that $E_3 > E_2 > E_1$
(Fig.1.9(a)). The atomic transition $|1\rangle \rightarrow |2\rangle$, with transition frequency
ω_{21}, interacts with the probe laser having frequency ω_P. The other atomic
transition $|2\rangle \leftrightarrow |3\rangle$, with transition frequency ω_{23}, interacts with the cou-
pling laser of frequency ω_C. The laser frequency detunings for the probe
and coupling transitions are $\Delta_P = \omega_P - \omega_{21}$ and $\Delta_C = \omega_C - \omega_{23}$, respectively. The semiclassical Liouville equation of the density-matrix elements
for this ladder-type three-level system under the dipole and RWA are given
by [4,24]

$$\dot{\rho}_{11} = \Gamma_2\rho_{22} + i\mu_{21}E_P\rho_{12} - i\mu_{21}E_P^*\rho_{21},$$
$$\dot{\rho}_{22} = \Gamma_3\rho_{33} - \Gamma_2\rho_{22} + i\mu_{32}E_C\rho_{23} - i\mu_{32}E_C^*\rho_{32} - i\mu_{21}E_P\rho_{12} + i\mu_{21}E_P^*\rho_{21},$$
$$\dot{\rho}_{33} = -\Gamma_3\rho_{33} - i\mu_{32}E_C\rho_{23} + i\mu_{32}E_C^*\rho_{32},$$
$$\dot{\rho}_{32} = -(\gamma_{32} - i\Delta_C)\rho_{32} + i\mu_{32}E_C(\rho_{33} - \rho_{22}) + i\mu_{21}E_P^*\rho_{31},$$
$$\dot{\rho}_{21} = -(\gamma_{21} - i\Delta_P)\rho_{21} + i\mu_{21}E_P(\rho_{22} - \rho_{11}) - i\mu_{32}E_C^*\rho_{31},$$
$$\dot{\rho}_{31} = -(\gamma_{31} - i(\Delta_C + \Delta_P))\rho_{31} - i\mu_{32}E_C\rho_{21} + i\mu_{21}E_P\rho_{32}. \tag{1.50}$$

The diagonal elements of this closed system satisfy the relation $\rho_{11} + \rho_{22} + \rho_{33} = 1$, which comes because of the trace condition of the density operator for the three-level system. The quantities defining dipole moments for the probe and coupling transitions are given by μ_{21} and μ_{23}, respectively. Note that these elements are considered to be real just for the sake of simplicity. The amplitudes of the probe and coupling fields are given by E_P and E_C, respectively. The radiative decay rates for diagonal density operator elements are given by Γ_i ($i = 1, 2, 3$) but the ground state is assumed to be stable, so $\Gamma_1 = 0$. The off-diagonal elements of density operator dephase out with decay constants $\gamma_{ij} = (\Gamma_i + \Gamma_j)/2$. The origin of EIT comes from coherence term ρ_{31} which is coupled to ρ_{21} and ρ_{32}. The existence of coherence term ρ_{31} under steady-state condition is caused by the presence of the coupling field E_C, which will be discussed in the following [24].

The general solution of density-matrix equations can be obtained numerically if the atomic system interacts with laser pulses of finite durations. In this way the transient effects related to EIT [69] can be studied. The Maxwell's equations for field propagations are also needed along with the density-matrix element equations in order to understand propagations of the matched pulses [70], propagation losses and pulse shape modifications [71].

When cw lasers are utilized for the probe and coupling transitions in this three-level system, the steady-state limit is quite sufficient to provide all the information about the system in terms of its absorption, dispersion, and nonlinear properties. For example, in the steady-state it is possible to obtain [24]

$$\rho_{31} = -\frac{i\mu_{32}E_C}{\gamma_{31} - i(\Delta_P + \Delta_C)}\rho_{21} + \frac{i\mu_{21}E_P}{\gamma_{31} - i(\Delta_P + \Delta_C)}\rho_{32}$$
$$\simeq -\frac{i\mu_{32}E_C}{\gamma_{31} - i(\Delta_P + \Delta_C)}\rho_{21}. \tag{1.51}$$

In the above expression the term proportional to ρ_{32} is not retained since this term is multiplied by the weak probe field E_P so the resultant magnitude is quite small compared to the first term and therefore can be neglected. Also, the populations in levels $|2\rangle$ and $|3\rangle$ are negligibly small which means that the value of ρ_{32} is also small. It is also assumed that in steady state to the first order in E_P/E_C, $\rho_{11} \cong 1$, $\rho_{22} \cong \rho_{33} \cong 0$. The probe coherence term ρ_{21} in the steady-state then reads as [4,24]

$$\rho_{21} = -\frac{i\mu_{21}E_P}{\gamma_{21} - i\Delta_P + \frac{\Omega_C^2/4}{\gamma_{31} - i(\Delta_P + \Delta_C)}}, \tag{1.52}$$

where $\Omega_C = 2\mu_{32}E_C$ is the Rabi frequency of the coupling field under the assumption that amplitude of the coupling field is real. The polarization P and complex susceptibility χ of the medium at the probe field frequency is [24]

$$P = \frac{1}{2}\varepsilon_0 E_P[\chi(\omega_P)e^{-i\omega_P t} + c.c.] = -2\mu_{21}N\rho_{21}e^{-i\omega_P t} + c.c., \tag{1.53}$$

such that

$$\chi = \frac{4i\mu_{21}^2 N/\varepsilon_0}{\gamma_{21} - i\Delta_P + \frac{\Omega_C^2/4}{\gamma_{31} - i(\Delta_P + \Delta_C)}}, \tag{1.54}$$

where N is the atomic density. The complex susceptibility can be written as $\chi = \chi' + i\chi''$, in which real (χ') and imaginary (χ'') parts are responsible for dispersion and absorption characteristics of the medium, respectively. The coefficients related to the intensity absorption and dispersion are characterized by parameters $\alpha = \omega_P n_0 \chi''/c$ and $\beta = \omega_P n_0 \chi'/2c$, respectively, and n_0 is the refractive index of the background. The Doppler broadening, which is a prominent broadening mechanism of absorption lines in atomic vapor can easily be included in the above expression of χ [24]. In the ladder-type EIT system under consideration, the probe and the coupling laser beams are propagating in opposite directions to each other, i.e., they are assumed to counter-propagate and also they have almost similar frequencies. When an atom moves towards the probe laser beam with velocity v, it sees blue-shifted laser frequency from the probe laser. Since the coupling laser is moving in the opposite direction to the probe laser so atom is moving away from the coupling laser beam and hence the red-shifted coupling laser frequency is seen by the atom. Alternatively, the frequency detuning of the probe laser will be blue-shifted to $\Delta_P + \omega_P v/c$ and that of the coupling laser will be red-shifted to $\Delta_C - \omega_C v/c$. The atomic density can be obtained using the Maxwell-Boltzmann distribution function $N(v)dv = (N_0/u\sqrt{\pi})\ \exp(-v^2/u^2)dv$ in which parameter u is a function of temperature T, given by $u = \sqrt{2k_B T/M}$, where M is the mass of the atom (see Eq.(1.31)). For a medium where line broadening is governed by Doppler profile, the FWHM of the absorption profile is $\Delta\omega_D = 2\omega_P(u/c)\ln(\sqrt{2})$. In a particular situation of $\omega_P \sim \omega_C$, it is possible to integrate expression of χ over all velocities to obtain [24]

$$\chi = \frac{4ic\mu_{21}^2 N_0 \sqrt{\pi}}{\varepsilon_0 u\omega_P}e^{z^2}(1 - erf z), \tag{1.55}$$

in which $erf(z)$ is the error function with the complex argument

$$z = \frac{c}{u\omega_P}[\gamma_{21} - i\Delta_P + \frac{\Omega_C^2/4}{\gamma_{31} - i(\Delta_P + \Delta_C)}]. \tag{1.56}$$

Since the probe laser and coupling lasers are counter-propagating in the ladder-type system, the first-order Doppler effect is canceled in the quantity $\Delta_P + \Delta_C$, and hence the EIT can be observed even with relatively weak coupling laser power [24].

If the probe and coupling laser beams undergoes collinear propagation in the ladder-type system, EIT effect will be completely wiped out in an atomic vapor cell, except with an extremely strong coupling beam (such as with pulsed laser) [21,22]. In such two-photon Doppler-free configuration only the first-order Doppler effect is eliminated. If the probe and coupling laser frequencies are different, higher-order Doppler effect still can affect EIT, as discussed in detail in Ref.[24]. The pioneering EIT experiment with cw diode lasers was demonstrated with rubidium atomic vapor at room temperature in three-level ladder-type configuration using D_2 line of ^{85}Rb [4,24]. In this experiment the coupling laser excites the upper transition from state $5P_{3/2}, F' = 4$ (level $|2\rangle$) to state $5D_{5/2}, F'' = 5$ (level $|3\rangle$) at the wavelength of 775.8 nm and the probe laser couples the lower transition $5S_{1/2}, F = 3$ (level $|1\rangle$) to $5P_{3/2}, F' = 4$ (level $|2\rangle$) at the wavelength of 780 nm. The three-level system used here and the one described in the theoretical model above are identical with $\Gamma_2 = 6.0$ MHz, $\Gamma_3 = 0.97$ MHz, and the Doppler width of about 540 MHz at room temperature. In the original EIT experiment with cw lasers, the probe and coupling laser beams were provided by two home-built diode lasers which were both temperature and current stabilized with free-running linewidth of about 5 MHz (not locked to FP cavity or atomic transition). Also, the probe and the coupling beams were orthogonally polarized and counter-propagating through the rubidium vapor cell.

In absence of the coupling beam, a normal absorption profile is obtained for the two-level transition from state $5S_{1/2}, F = 3$ to state $5P_{3/2}, F' = 4$, as shown in Fig.1.10(a). The measured maximum absorption at the center of the Doppler-broadened line was found to be $\alpha = 8.2 \times 10^{-2}$ cm^{-1} at $T = 21$ ^0C. After turning on the coupling field at the resonance frequency, i.e., $\Delta_C = 0$, a narrow dip at the center of the absorption profile (Fig.1.10(b)) comes out. The modified absorption coefficient measured in this condition was $\alpha = 2.9 \times 10^{-2}$ cm^{-1} at the central frequency. So, there was an absorption reduction of 64.4% due to the induced atomic coherence created by the coupling field. The experimental results agree quite well with the theoretical calculation shown in Figs.1.10(a) and 1.10(b) with only the coupling field Rabi frequency as a fitting parameter. In Fig.1.10(b) the coupling Rabi frequency is estimated to be $\Omega_C = 92$ MHz, which is much smaller than the Doppler width in this system. This clearly brings out the advantage of using two-photon Doppler-free configuration in such EIT system. The prime reason for the limited absorption reduction (EIT) in this three-level atomic system in the ladder-type configuration is due to the high dephasing rate γ_{31} and the relatively broad free-running linewidths of the diode lasers used in the experiment [4,24].

When the coupling laser beam is not on resonance with its transition frequency, a dispersive-like feature is observed at the side of the Doppler-broadened absorption profile. Such a feature is due to the admixture of two effects, i.e., absorption reduction due to atomic coherence and enhancement due to two-photon absorption [24]. Another observation in the experiment was the absorption peak at far detuning of the coupling laser beam, meaning Δ_C goes past beyond Doppler profile. This peak absorption was due to two-photon absorption from state $|1\rangle$ to state $|3\rangle$. These observed phenomena can all be well explained by the theoretical results of Eq.(1.55) [24]. It is possible to control probe beam absorption [24,72] by adjusting the

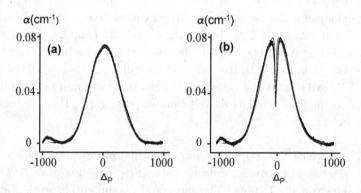

Fig. 1.10   Absorption coefficient α as a function of probe detuning Δ_P. Heavy black curve, experiment; gray curve, theory. (a) No coupling laser. (b) Coupling laser on resonance $\Delta_C = 0$. The theoretical parameters are $\gamma_{21} = 3$ MHz, $\gamma_{31} = 0.5$ MHz, $\Omega_c = 92$ MHz, $\Delta\omega_D = 540$ MHz, $\delta\omega_D = -2.97$ MHz, and laser (half) linewidth $\gamma_C = \gamma_P = 2.5$ MHz (adopted from [24] with permission).

coupling beam intensity at $\Delta_C = 0$ (Fig.1.11(a)) or $\Delta_C \neq 0$ (Fig.1.11(b)). When power of the coupling laser beam is increased, there is a decrease in the absorption coefficient of the probe transition measured under resonance condition, i.e., at the central frequency of the probe laser beam. Such a absorption reduction is the EIT behavior of the system and is due to the atomic coherence induced by the coupling beam as shown in Fig.1.11(a). For off-resonance condition ($\Delta_C \cong -550$ MHz) of the coupling laser, the two-photon absorption coefficient goes up with the increase of coupling beam power (Fig.1.11(b)).

The dispersion properties of the probe field in such EIT system is also controllable, as can be easily seen from the real part of Eq.(1.55). Such modified dispersive properties of the three-level atomic medium for the probe laser beam was first measured using a Mach-Zehnder interferometer in the same atomic system used above for demonstrating EIT [25]. The small phase shift $\beta(\omega)L$ for the probe beam introduced by the atoms in the atomic vapor cell was detected through a setup called homodyne detection system. The differential intensity for the signal (probe) of the balanced homodyne detectors is given by [25]

$$\Delta I_d(\omega) \propto 2|E_{LO}||E_S|e^{-\alpha(\omega)L/2}\cos[\phi_{LO} + \beta(\omega)L], \qquad (1.57)$$

in which the quantities E_S and E_{LO} represent the signal field and local oscillator field passing through the cell and reference arm, respectively, of

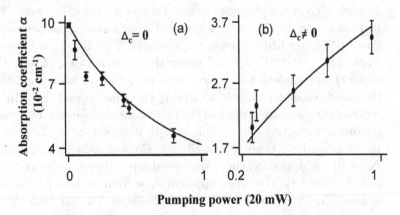

Fig. 1.11 Absorption coefficient α as a function of coupling (pumping) beam power for (a) $\Delta_C = \Delta_P = 0$ and (b) $\Delta_C = -550$ MHz and $\Delta_P \sim 550$ MHz (the two-photon absorption peak is slightly shifted from $\Delta_P = -\Delta_C$). Solid lines, theoretical predictions and parameters are the same as in Fig.1.10 (adopted from Fig.7 of [24] with permission).

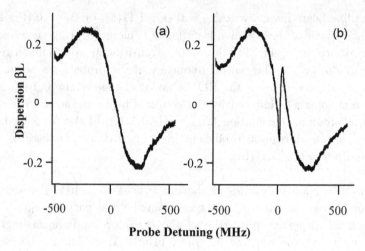

Fig. 1.12   Measured dispersion βL of the rubidium vapor vs probe detuning Δ_P. (a) dispersion with no coupling laser; (b) dispersion with coupling laser on resonance $\Delta_C = 0$ (adopted from [25] with permission).

Mach-Zehnder interferometer [4,25]. In the experiment the intensity of the signal laser beam was kept far below the rubidium D_2 transition saturation level and the condition $|E_S| << |E_{LO}|$ was also maintained. The parameters used in the experiments were defined as follows : $\alpha(\omega)$ is the absorption coefficient, $\beta(\omega)$ is the dispersion coefficient, L is the length of the rubidium cell, ϕ_{LO} is the phase of the interferometer's reference beam. For each frequency scan, the phase ϕ_{LO} was reset to $\pi/2$ using a piezoelectric transducer (PZT) on which a mirror is mounted [25]. Hence, it is possible to write $\Delta I_d \propto e^{-[\alpha(\omega)L/2]}\beta(\omega)L$ under the condition $|\beta(\omega)|L << 1$. Time required to complete a frequency scan of the probe laser was 50 ms and the interferometer had no drift during that time period [25]. In the absence of the coupling beam a typically measured dispersion curve of probe transition for a two-level atomic system is shown in Fig.1.12(a). When the coupling laser beam is turned on under the resonance condition (i.e., $\Delta_C = 0$) the dispersion curve shows a sharp variation near $\Delta_P = 0$ as depicted in Fig.1.12(b) with an opposite slope from the case for the two-level system. This curve provides the change in dispersion near center frequency, which was found to be $d(\beta L)/d\nu|_{\nu_0} \cong 19.4 \times 10^{-9}$ sec. The estimated group velocity corresponding to this dispersion change is $v_g = c/13.2$. The slowing down of the group velocity in the probe laser beam can be attributed to the rapid variation in refractive index or the enhanced normal dispersion due to the coupling beam-induced atomic coherence in this EIT system.

Significant improvements in experimental conditions could and had been made over the years from this initial experiment and further reduction of group velocity has been observed. For example, with narrow linewidth and frequency-locked laser sources, specially chosen atomic levels, and inclusion of buffer gases in the atomic vapor cell, the dispersion change can be greatly enhanced. Later, with those improvements made using a Λ-type system of hot rubidium atomic vapor the group velocity of light was controlled by the coupling laser beam and slowed down to 90 m/s [73]. With a cold atomic sample, the group velocity of a probe pulse was reduced down to even 17 m/s [74].

B. Three-level system in Λ-type configuration

Next, we consider another interesting three-level configuration known as Λ-type system. This system is shown in Fig.1.9(b), where the probe (coupling) laser beam with frequency ω_P (ω_C) connects the transition from level $|2\rangle$ to level $|1\rangle$ ($|3\rangle$). The Liouville equation of the density-matrix elements for this system in dipole and RWA is given by [4,24]

$$\dot{\rho}_{11} = \gamma_{31}(\rho_{33} - \rho_{11}) + \gamma_{21}\rho_{22} - \frac{i}{2}\Omega_P^*\rho_{21} + \frac{i}{2}\Omega_P\rho_{12},$$

$$\dot{\rho}_{22} = -(\gamma_{23} + \gamma_{21})\rho_{22} + \frac{i}{2}\Omega_P^*\rho_{21} - \frac{i}{2}\Omega_P\rho_{12} + \frac{i}{2}\Omega_C^*\rho_{23} - \frac{i}{2}\Omega_C\rho_{32},$$

$$\dot{\rho}_{33} = \gamma_{31}(\rho_{11} - \rho_{33}) + \gamma_{23}\rho_{22} - \frac{i}{2}\Omega_C^*\rho_{23} + \frac{i}{2}\Omega_C\rho_{32},$$

$$\dot{\rho}_{23} = -(\frac{\gamma_{21} + \gamma_{32} + \gamma_{31}}{2} - i\Delta_C)\rho_{23} + \frac{i}{2}\Omega_C(\rho_{22} - \rho_{33}) - \frac{i}{2}\Omega_P\rho_{13},$$

$$\dot{\rho}_{21} = -(\frac{\gamma_{21} + \gamma_{32} + \gamma_{31}}{2} - i\Delta_P)\rho_{21} + \frac{i}{2}\Omega_P(\rho_{22} - \rho_{11}) - \frac{i}{2}\Omega_C\rho_{31},$$

$$\dot{\rho}_{31} = -(\gamma_{31} - i(\Delta_P - \Delta_C))\rho_{31} - \frac{i}{2}\Omega_C^*\rho_{21} + \frac{i}{2}\Omega_P\rho_{32}. \tag{1.58}$$

in which Ω_P and Ω_C are Rabi frequencies of the probe and coupling fields, respectively defined as $\Omega_P = 2\mu_{21}E_P/\hbar$, $\Omega_C = 2\mu_{32}E_C/\hbar$, assuming amplitudes of the probe and coupling fields to be real. The radiative decay constants are defined as follows, e.g., γ_{21} is the radiative decay constant for level $|2\rangle$ to $|1\rangle$ and γ_{23} is the radiative decay constant for level $|2\rangle$ to $|3\rangle$. The nonradiative decay constant connecting levels $|3\rangle$ and $|1\rangle$ is γ_{31} and $\gamma = (\gamma_{21} + \gamma_{23} + \gamma_{31})/2$. Here Δ_P and Δ_C are the frequency detunings of the probe and coupling lasers, respectively (as defined before Eq.(1.50)).

When the EIT condition is satisfied for the system meaning the coupling field is much stronger than the probe field, i.e., $\Omega_C >> \Omega_P$, then in the steady-state condition, almost all the atoms occupy level $|1\rangle$ due to optical pumping by the strong coupling beam and hence $\rho_{11} \cong 1$, $\rho_{22} \cong \rho_{33} \cong 0$ (within the first-order approximation in Ω_P/Ω_C). The requirement of Doppler-free condition for such Λ-type system is the collinear propagation of the probe and coupling laser beams. By employing such co-propagating configuration the first-order Doppler shifts of the probe laser and the coupling laser for the same group of atoms with a particular velocity can cancel each other [4,24]. As discussed above for the ladder-type EIT system, the Doppler broadening for the Λ-type EIT system can also be worked out easily by integrating the expression of χ over the Maxwell-Boltzmann velocity distribution. Hence, in the steady state the susceptibility is given by the expression [24]

$$\chi = \frac{4ic\mu_{21}^2 N_0 \sqrt{\pi}}{\epsilon_0 u \omega_P} e^{z^2}(1 - erf(z)), \tag{1.59}$$

with

$$z = \frac{c}{u \omega_P}[\gamma - i\Delta_P + \frac{\Omega_C^2/4}{\gamma_{31} - i(\Delta_P - \Delta_C)}]. \tag{1.60}$$

To observe EIT in such Λ-type system, a bit different arrangement is required from the ladder-type system. In the experimental setup for observing EIT in Λ-type system, collinear propagation of the probe and coupling laser beams is required in order to eliminate first-order Doppler effect. The polarizations of the probe and coupling lasers are kept orthogonal to each other and hence a polarization cube beam splitter can be used to combine these two beams before they enter the rubidium atomic vapor cell. Similarly another polarization cube beam splitter is used after the rubidium cell to separate these two beams from each other [23]. To observe EIT, different polarizations for the probe and coupling lasers are not a stringent criterion. In the original experiment two different polarizations were used so that two beams can be easily combines at the input and separated physically at the time of detection.

The three-level Λ-type system can be realized in ^{87}Rb by considering the D_1 line transitions as shown in Fig.1.9(b) [23]. The two lower levels $|1\rangle$ and $|3\rangle$ of the Λ-type system are hyperfine levels $F = 1$ and $F = 2$, respectively, of the ground state $5S_{1/2}$. The hyperfine level $F' = 2$ of $5P_{1/2}$ serves as the upper level $|2>$. Another hyperfine level $F' = 1$ of the excited state $5P_{1/2}$ is 812 MHz away, which does not participate directly in this configuration because it lies outside the Doppler width of the transition lines. The interaction of the probe laser is with $5S_{1/2}$, $F = 1$ and $5P_{1/2}$, $F' = 2$ states and the associated Rabi frequency is Ω_P, while the coupling laser is tuned to interact with $5S_{1/2}$, $F = 2$ and $5P_{1/2}$, $F' = 2$ states and the Rabi frequency is Ω_C.

The measured absorption profiles of $5S_{1/2}$, $F = 1 \rightarrow 5P_{1/2}$, $F' = 2$ transition with and without the presence of the coupling beam are shown in Fig.1.13. The lower solid curve with no dip in this figure represents the situation when the coupling laser is absent. In the presence of the coupling field, the population is pumped from level $|3\rangle$ and eventually it goes down to level $|1\rangle$ by the strong coupling beam. Thus the total absorption of the probe beam should increase but with a large dip at the center of line profile due to EIT, which is clearly represented by the upper solid curve. The peak absorption coefficients in the absence and presence of the coupling field were found to be $\alpha = 0.92 \times 10^{-2}$ cm^{-1} and $\alpha = 0.56 \times 10^{-2}$ cm^{-1}, respectively. By taking into account the absorption increase due to optical pumping ($\alpha = 1.92 \times 10^{-2}$ cm^{-1}), the absorption reduction was found to be 70.8% with respect to the new absorption peak [4,23]. The dip goes down below the unpumped level as shown in Fig.1.13 was a clear demonstration of atomic coherence generated in this EIT system. There was a good agreement between theoretically fitted and experimentally measured results. When the coupling laser was tuned to interact with $5S_{1/2}$, $F = 2 \rightarrow 5P_{1/2}$, $F' = 1$ transition, the EIT effect was also observed in $5S_{1/2}$, $F = 1 \rightarrow 5P_{1/2}$, $F' = 1$ transition. When the pumping intensity was 560 mW/cm^{-2}, the absorption reduction was measured to be about 85.4% [23]. The initial experimental demonstrations of EIT, especially with cw diode lasers [4,23,24], have started many new research activities, e.g., absorption reduction of more than 90% in atomic vapor cells and cold atomic samples were reported subsequently. Doubly dressed states were observed experimentally in ^{87}Rb atoms cooled and confined in a magneto-optical trap [4,75]. Such doubly-dressed states were obtained by a combination of a strong-coupling laser and a moderate intensity pump laser in a three-level atomic system. A three-peak spectral profile obtained in the absorption spectrum of a weak probe laser could be interpreted using the dressed-state picture. This interpretation agrees quite well with the results obtained by the density-matrix approach.

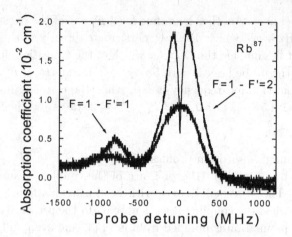

Fig. 1.13   Absorption coefficients for the probe beam versus probe frequency for the $5S_{1/2}, F = 1 \rightarrow 5P_{1/2}, F' = 2$ transition of ^{87}Rb. The lower solid curve is for no coupling field. The upper solid curve is for the coupling field tuned at the $5S_{1/2}, F = 2 \rightarrow 5P_{1/2}, F' = 2$ transition with intensity $I_c = 19.6$ W cm^{-2} at the cell center (adopted from [23] with permission).

These experimental demonstration have clearly revealed that the linear absorption and dispersion properties of the probe beam can be conveniently altered and controlled in the three-level atomic systems due to the coupling beam-induced atomic coherence. Such controllability is also very useful in modifying nonlinear optical properties of such three-level EIT system, which will be discussed in the following.

1.5.4   *Enhancement and control of Kerr nonlinearity in three-level electromagnetically induced transparency systems*

From the above discussion in subsection 1.5.3, it is quite clear that the atomic coherence induced by the coupling and probe laser beams is responsible to change absorption and dispersion properties of the probe beam. The steepness of dispersion slope so observed under EIT condition is responsible for the substantial reduction in the group velocity of probe pulses [25,73,74]. This reduction in group velocity brings a significant increase in the effective interaction time for the probe pulses with the atomic medium. Optical nonlinearities in multilevel systems could be enhanced by the induced atomic coherence and thus facilitates to realize nonlinear optical processes using ultra weak optical power levels.

Usually, if the nonlinear optical susceptibility is resonantly enhanced at very near transition resonances, the medium will show very large near-resonant absorption and hence its opacity increases. However, in an EIT setup there is a reduced absorbtion under resonant condition and, at the same time, there is great enhancement in nonlinear susceptibility due to the presence of induced atomic coherence in such multilevel systems [4]. One model to create a nonlinear medium using a multilevel system was suggested in Ref.[67]. The level configuration used in that model is shown in Fig.1.14. Here a strong coupling laser beam of frequency ω_C interacts with two metastable states $|2\rangle$ and $|3\rangle$. The dipole forbidden transition between levels $|1\rangle$ and $|2\rangle$ can be excited with a two-photon transition $(\omega_a + \omega_b)$ to generate a sum frequency $\omega_d = \omega_a + \omega_b + \omega_c$. The resonance transition $|1\rangle \rightarrow |3\rangle$ shows strong absorption at ω_d if the field at ω_c is not present. The medium shows transparency at the line center when the Rabi frequency of the coupling field is much larger compared to the Doppler width of the $|1\rangle \rightarrow |3\rangle$ transition [4]. Such transparency is a result of the destructive interference between the Autler-Townes split components of the $|1\rangle \rightarrow |3\rangle$ transition created by the coupling field at ω_C. This interference does not reduce the nonlinearity of the system responsible for the generation of ω_d, due to a change of sign in the dressed eigenvector. As a matter of fact, there is a constructive rather than destructive interference in the nonlinear susceptibility [4,67] in view of the up-converted frequency lying between the Autler-Townes components. This example of the multilevel system is unique in the sense that it predicts the role of induced atomic coherence in the reduction of linear absorption and simultaneously enhancing the nonlinearity.

In the past fifteen years there have been quite many reports on the experimental demonstrations of enhanced nonlinear optical processes in a variety of multilevel atomic systems [4]. Since atomic systems have central symmetry, the second-order nonlinear susceptibility is typically zero, so we will only consider the third-order (Kerr) nonlinearity in the following studies. There are generally two kinds of Kerr nonlinearities, e.g., the cross-Kerr nonlinearity and self-Kerr nonlinearity. The cross-Kerr nonlinearity is the nonlinear index change of one optical beam caused by another beam, and self-Kerr nonlinearity is the nonlinear index change caused by the beam itself. The enhanced multi-wave mixing processes are due to the cross-Kerr

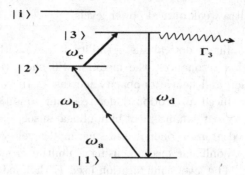

Fig. 1.14 Schematics of energy levels for the sum-frequency generation process $\omega_d = \omega_a + \omega_b + \omega_c$. The state $|3>$ has spontaneous decay rate Γ_3. When a strong field at frequency ω_C is tuned to the line center of the $|2\rangle \rightarrow |3\rangle$ transition, the medium becomes transparent on the $|1\rangle \rightarrow |3\rangle$ transition for the generated field (adopted from [67] with permission).

nonlinearities, but the AOB phenomena are typically the results of the self-Kerr nonlinear index of refraction. So, in the following, we will mainly concentrate on the self-Kerr nonlinearity. Experimental demonstrations of enhanced nonlinear optical processes can be realized easily but direct measurements of the Kerr nonlinear coefficients associated with various nonlinear processes are not so easy because the residual linear absorption and dispersion effects can not be eliminated completely. The normal techniques of measuring self-Kerr nonlinearity, such as Z-scan, do not provide proper results when long vapor cells are used for confining the atomic medium. The effects arising due to enhanced Kerr nonlinearity in multilevel systems are useful in many applications including self-phase modulation for optical shutters, four-wave mixing process for frequency conversion, logic gate implementations, and entangled states for quantum information processing [4]. Doing direct measurement of the Kerr-nonlinear coefficient can provide good understanding of nonlinear optical processes and their effective controls using experimental parameters of the system.

Direct measurement of the self-Kerr nonlinear index of refraction has been experimentally demonstrated by using atomic vapor cell placed inside an optical cavity and then measuring the nonlinear phase shifts on the cavity output profile, which are not affected by linear optical properties of the atomic medium, as can be seen from the cavity transmission function given by Eqs.(1.48) and (1.49) and other details in Ref.[33,76]. The nonlinear susceptibility for the probe beam can be easily calculated for a three-level atomic medium using the density-matrix equations as shown in subsections 1.2.3 and 1.5.3. The methodology to obtain nonlinear susceptibility is quite similar to the one used for calculating the linear susceptibility (in subsections 1.2.3 and 1.5.3), however, it is required to keep the probe intensity to higher order (for the Kerr-nonlinearity this is up to the third order) by using the iterative technique [33,76]. After some calculation, one gets the following expression for the total susceptibility of the probe transition in the three-level Λ-type system as [4,33,76]

$$\chi \cong \frac{iN|\mu_{21}|^2}{\hbar} \frac{1}{F} [1 - \frac{2\gamma_{31}}{2\gamma + \gamma_{21}} - \frac{|\Omega_P|^2}{2\gamma + \gamma_{21}} \frac{F + F^*}{|F|^2}], \qquad (1.61)$$

with $F \equiv \gamma - i\Delta_P + (|\Omega_C|^2/4)/[\gamma_{31} - i(\Delta_P - \Delta_C)]$. In Eq.(1.61), the first term is the linear susceptibility, the second term is the contribution to the linear susceptibility from the second order term in the expansion of $\rho_{11} - \rho_{22}$ [76], and the third term is related to the self-Kerr third-order nonlinearity $\chi^{(3)}$ due to the probe intensity appearing in the expression: $\chi \cong \chi^{(1)} + 3\chi^{(3)}|E_P|^2$ with the modification provided by atomic coherence. It is then easy to obtain the expression for the self-Kerr nonlinear index of refraction n_2 [3,4,33,76]

$$n_2 = \frac{12\pi^2}{n_0^2 c} Re\chi^{(3)},$$

$$= Re[-\frac{4iN\pi^2|\mu_{21}|^4}{\hbar^3 c n_0^2} \frac{1}{2\gamma + \gamma_{21}} \frac{F + F^*}{F|F|^2}]. \qquad (1.62)$$

For thermal atoms moving in a vapor cell, the Doppler effect can be taken into account by integrating Eq.(1.62) over a Maxwell-Boltzmann velocity distribution as discussed above for the linear susceptibility in section 1.5.3. The removal of first-order Doppler effect is more or less achieved by co-propagating the coupling and probe laser beams through the atomic cell (assuming the Λ-type configuration is used for the atoms) kept inside the optical ring cavity. Considering there is no Doppler effect and γ_{31} negligible, then from the expression given in Eq.(1.62), the maxima of n_2 are obtained at $\Delta_P \cong \pm\Omega_C/2$ for $\Delta_C = 0$. It can be seen that $n_2 > 0$ for $\Delta_P = -\Omega_C/2$ and $n_2 < 0$ for $\Delta_P = \Omega_C/2$. In yet another situation, if $\Delta_P = 0$ and there is variation in Δ_C then n_2 behaves in opposite way. This is because both Δ_C and Δ_P behave in a similar way on n_2 but their signs are opposite [4,33].

The required setup for experimental measurement of n_2 in the three-level Λ-type EIT system is sketched in Fig.1.15. A 5 cm long cell with Brewster windows was containing the rubidium atomic vapor, which was wrapped in μ-metal sheets to shield it from magnetic field. The temperature of the cell was kept constant with a controlled heater to about $63\,^{\circ}$C. This arrangement of vapor cell was placed inside a three mirrored optical ring cavity. The mirrors M2 and M3 (both concave with radius of curvature $R = 10$ cm) had reflectivities of 97% and 99.5%, respectively and the mirror M3 was mounted on a PZT. On the other hand the mirror M1 (the plane mirror) of this cavity had reflectivity of 99%. The measured Finesse (F) of the optical cavity was about 55 when Rb vapor cell was kept in it. The measurement of F was performed far from any resonant absorption line of rubidium. The complete length of cavity in the experiment was about ~ 37 cm, which corresponds to a FSR of about 822 MHz. During the experiments the probe laser beam was injected through mirror M2 and it circulated in the cavity in a single direction. The coupling laser beam entered in the cavity through a polarizing beam cube with an orthogonal polarization with respect to the probe beam and did not circulate in the cavity (Fig.1.15). It is easy to measure frequency detunings Δ_C and Δ_P with the help of another FP cavity in conjunction with a saturation absorption spectroscopy (SAS) setup [4,33]. To begin with, the coupling laser was tuned and locked to the transition $5S_{1/2}, F = 2 \rightarrow 5P_{1/2}, F' = 2$, e.g., $\Delta_C = 0$ of ^{87}Rb atom as shown in Fig.1.9(b). After that the probe beam was tuned and locked to the transition $5S_{1/2}, F = 1 \rightarrow 5P_{1/2}, F' = 2$ slightly below or above resonance. The estimated values of average Rabi frequencies related to the coupling and the probe laser fields were $\Omega_C = 2\pi \times 72$ MHz and $\Omega_P = 2\pi \times 11$ MHz, respectively. The optical cavity length was scanned by the mirror mounted on PZT by applying a suitable scanning voltage on PZT and the transmission peaks of cavity were monitored by an avalanche photodiode detector (APD).

When there was no coupling field present in the cavity ($\Omega_C = 0$, no EIT), the typical cavity transmission profile was a symmetric one. In the presence of coupling beam and also EIT , the cavity transmission profile became asymmetric due to the nonlinearity in the phase shift caused by intracavity medium. Measuring the degree of asymmetry in the cavity transmission profile can provide a direct measure of nonlinear phase shift which is directly related to n_2 and given by [33]

$$\delta = \frac{2\pi[(n_0 - 1)L + L_C]}{\lambda} + \frac{2\pi L}{\lambda} n_2 I_P + \Phi_0 - 2\pi m, \qquad (1.63)$$

in which L is the length of the atomic vapor medium, L_C is the length of optical cavity, m is an integer, λ is the wave-length of the cavity field, Φ_0 is the phase offset related to the cavity, and I_P is the intracavity probe beam intensity. Basically this expression is similar to Eq.(1.49) with some rearrangements of terms.

The sign of the Kerr nonlinear coefficient n_2 is determined by knowing in which direction the asymmetry of output transmission peak projects out. When the cavity scan is performed from smaller to bigger length, the intracavity intensity requires longer to achieve the peak value for negative value of n_2, because the sign of second term in Eq.(1.63) is opposite to that of the first term. For the positive value of n_2, the asymmetry in cavity transmission profile gets reversed. The measured value of Kerr nonlinear index of refraction n_2 can be easily calculated using the following expression [33,76]

$$n_2 = \lambda \frac{[(t_1 - t_r) - (t_r - t_2)]}{2\tau(I_r - I_\delta)L}, \tag{1.64}$$

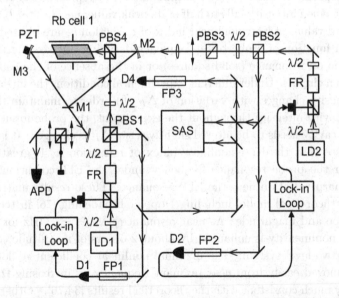

Fig. 1.15 Experimental setup. LD1 and LD2 are coupling and probe diode lasers, respectively; PBS1-PBS4 are polarizing cubic beam splitters; $\lambda/2$, half-wave plates; FR, Faraday rotators; FP1-FP3, Fabry-Perot cavities; D1-D4, detectors; SAS is the unit for saturation absorption spectroscopy; APD, avalanche photodiode detector (reprinted from [33] with permission).

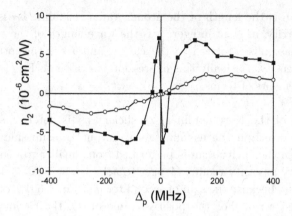

Fig. 1.16   Measured Kerr nonlinear coefficient n_2 versus frequency detuning of the probe beam with $\Delta_c = 0$ and $\Omega_c = 2\pi \times 72$ MHz. Solid squares are with coupling beam and open circles are without coupling beam (reprinted from [33] with permission).

in which I_r represents the maximum intensity value of the cavity transmission profile at time t_r. The parameters t_1, t_2 represent the values of scan time when intensity falls to half of its peak value, i.e., $I_r/2(\sim I_\delta)$. The measured values of Kerr nonlinear index of refraction n_2 are plotted in Fig.1.16 as a function of probe beam frequency detuning but keeping the coupling beam on resonance (solid squares) or in the absence of the coupling beam (open circles). Under different experimental conditions the intracavity peak intensity changes with variation of Δ_P. In order to maintain the same intracavity intensity throughout the experiment, the probe input power into the cavity needs to be altered slightly when Δ_P is changed. It is quite clear to see that the Kerr-nonlinear index of refraction n_2 is greatly modified near the probe resonance frequency under the EIT condition and it has a sharp variation near the EIT resonance. These results match well with the theoretical model including Doppler broadening [76] in terms of both shape and magnitude. At near resonance ($\Delta_P \cong \pm 7$ MHz for given Ω_C), the nonlinearity is enhanced by about 2 orders of magnitude compared to the two-level system. The maximal nonlinear coefficient n_2 locations can be moved easily from near resonant frequency by increasing Ω_C, which is very much consistent with the theoretical results [34,76]. Other interesting results for n_2 were obtained when coupling field detuning Δ_C was varied keeping $\Delta_P = 0$ as shown in Fig.1.17. The prediction of such observation can be made with Eq.(1.62) as the roles of Δ_P and Δ_C are different because their signs are different. This result is very useful in the sense that

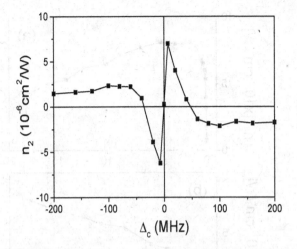

Fig. 1.17   Measured Kerr nonlinear coefficient n_2 versus frequency detuning of the coupling beam with $\Delta_p = 0$ and $\Omega_c = 2\pi \times 72$ MHz (reprinted from [33] with permission).

the sign of n_2 can be changed in a controlled manner by simply modifying Δ_C, i.e., the frequency of the coupling (or controlling) field, within a very small range [4,33].

Measurements of n_2 was also done by changing the coupling beam power P_C [34] under the parametric condition $\Delta_C = 0$, $\Omega_P = 2\pi \times 11$ MHz and for a fixed value of Δ_P. The plot of variation of n_2 with respect to P_C is provided in Fig.1.18(a) for $\Delta_P = 40$ MHz. The raise in Kerr nonlinearity is very rapid as the coupling power increases but it goes down slowly after reaching to a peak value. The behavior of n_2 is different at near resonance condition. For example, it was observed at $\Delta_P = 7$ MHz, n_2 keeps going up as the coupling power goes up, as shown in Fig. 1.18(b), and is limited only by the availability of coupling power in that experiment [4,34]. It should be noted that n_2 has a positive value for $\Delta_P = 40$ MHz but it becomes negative for $\Delta_P = 7$ MHz, as shown in Fig.1.16. The measured value of n_2 (7×10^{-6} cm^2/W) in these experiments is quite large compared to the possible attainable value in an usual two-level atomic vapor medium. The enhanced nonlinearity in the probe transition is because of the induced atomic coherence in the presence of the coupling laser beam. Thus, the measurements of n_2 with respect to the parameters Δ_P, Δ_C, P_C etc show that it is easy to manipulate and control the Kerr nonlinear coefficient of the probe transition by making changes in the experimental parameters

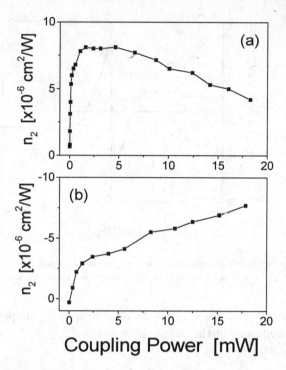

Fig. 1.18   Measured Kerr nonlinear index of refraction n_2 as a function of coupling beam power for (a) $\Delta_P = 40$ MHz and (b) $\Delta_P = 7$ MHz. Other parameters are $\Omega_P = 2\pi \times 11$ MHz and $\Delta_c = 0$ (reprinted from [34] with permission).

such as frequency detuning and power level of the coupling beam.

Note that the nonlinear index of refraction n_2, measured in this experiment is related to the term $n_2 I_P$, i.e., the intensity-dependent refractive index, which arises from the Kerr nonlinear susceptibility $\chi^{(3)}$ and give rise to the phenomenon of self-phase modulation. In the literature we also find n_2 is related to the expression $n_2' I_C$, which is determined from the slope of linear dispersion and creates cross-phase modulation (XPM) [73,77]. Controlling of self-focusing and self-defocusing were also reported earlier [78]. In a recent experiment results related to the direct measurement of cross-phase nonlinear refractive index n_2' were obtained in a three-level ^{87}Rb atomic system with the same method as discussed above [79].

Chapter 2

Atomic Optical Bistability in a Two-level System

2.1 Two-level Atoms inside an Optical Cavity

The quantum and nonlinear optical phenomena in homogeneously-broadened two-level atomic system confined inside an optical cavity are modeled extensively over the past three decades. Many such phenomena including optical bistabilty could be easily understood by the standard model given by Bonifacio and Lugiato [1,80]. The model used in Refs.[1-3,80] contains a unidirectional ring cavity, as shown in Fig.2.1, having four mirrors $(M_i, i = 1, 4)$. The input and output mirrors M_1 and M_2 of this ring cavity have reflection and transmission coefficients R and T, respectively, such that $R + T = 1$. The reflectivities of mirrors M_3 and M_4 are assumed to be near 100 % to make analysis simple. The sample of length L contains the two-level atoms and for the sake of simplicity, the Doppler broadening is ignored in the following discussion.

The two-level atomic system is characterized with energies E_i (i=1,2) for the two energy levels $(E_2 > E_1)$, with atomic transition frequency ω_{21}, the laser exciting this transition has frequency ω_P with a frequency detuning $\Delta_P = \omega_{21} - \omega_P$, and μ_{21} is the dipole moment of the atomic transition. In the semiclassical formulation the Liouville equation of the density operator in the dipole and RWA is given (see Eq.(1.12) in Chapter 1 and note that we used here $V_{21} = -\mu_{21}E_P e^{-i\omega_P t}$ (Eq.(1.15)) as [1,2]

$$\dot{\rho}_{11} = \Gamma_{21}\rho_{22} - i\mu_{21}\hbar^{-1}E_P\rho_{12} + i\mu_{21}\hbar^{-1}E_P^*\rho_{21},$$
$$\dot{\rho}_{22} = -\Gamma_{21}\rho_{22} + i\mu_{21}\hbar^{-1}E_P\rho_{12} - i\mu_{21}\hbar^{-1}E_P^*\rho_{21},$$
$$\dot{\rho}_{21} = -(\gamma_{21} + i\Delta_P)\rho_{21} - i\mu_{21}\hbar^{-1}E_P(\rho_{22} - \rho_{11}).$$

$$(2.1)$$

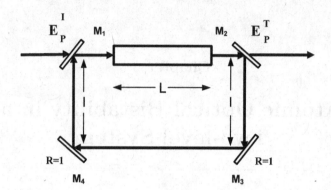

Fig. 2.1   A unidirectional ring cavity having four mirrors (M_1-M_4) and an atomic sample of length L. The mirrors M_3 and M_4 are perfectly reflecting mirrors ($R = 1$). The incident and the transmitted fields are represented by E_P^I and E_P^T, respectively (adopted from [4] with permission).

These density-matrix equations (Eq.(2.1)) describe the dynamical evolution of the two-level system. The electric field experienced by the atoms has the following form

$$\vec{E} = (\vec{E_P}e^{-i\omega_P t} + c.c.).\qquad(2.2)$$

This electric field, which oscillates at a frequency ω_P and interacts with the atomic transition $|1>$ to $|2>$, makes circular round trip inside the optical ring cavity and called as the cavity field. The atomic polarization induced in the medium giving rise to AOB is given by

$$P(\omega_P) = N\mu_{21}\rho_{12},\qquad(2.3)$$

in which the quantity N is the atomic number density. We can define the coupling constant as $g_{21} = \frac{4\pi\omega_P\mu_{21}}{V}$, in which V is the volume of atomic sample. The entry of field E_P into the optical cavity is through the partially transparent mirror M_1, which drives the two-level atomic transition. In the optical ring cavity, the incident field E_P^I, the transmitted field E_P^T and fields at different locations in the cavity ($E_P(0,t)$, $E_P(L,t)$) obey certain boundary conditions which are stated as follows [1,2,5]).

$$E_P^T = \sqrt{T}E_P(L,t),$$
$$E_P(0,t) = \sqrt{T}E_P^I(t) + Re^{-i\delta_0}E_P(L,t-\Delta t),\qquad(2.4)$$

where L is the atomic sample length and l = side arm length between M_2 (M_1) and M_3 (M_4) such that $\Delta t \cong (2l + L)/c$ is the time taken by light to travel from mirror M_2 to M_1 via M_3 and M_4. The frequency phase detuning of the cavity is defined as $\delta_0 = (\omega_{cav} - \omega_P)L_T/c$, in which ω_{cav} is the frequency of the cavity mode nearest to frequency ω_P, and $L_T \cong 2(l + L)$ amounts to the total length of the ring cavity.

Dynamics of the electric field inside the optical cavity can be obtained by using the Maxwell's equation in slowly-varying envelope approximations, i.e.,

$$\frac{\partial E_P}{\partial t} + c\frac{\partial E_P}{\partial z} = 2i\pi\omega_P\mu_{12}P(\omega_P), \tag{2.5}$$

and the circulating field obeys boundary conditions as defined in Eq.(2.4) above. To find the value of polarization $P(\omega_P)$, the solution of the density-matrix equations (2.1) in the steady-state condition is required, then Eq.(2.4) is used to integrate Eq.(2.5) in the steady-state limit over the length of the atomic sample to obtain the field. The field boundary conditions can be written in the steady-state as [1,5]

$$E_P^T = \sqrt{T}E_P(L),$$
$$E_P(0) = \sqrt{T}E_P^I + Re^{-i\delta_0}E_P(L). \tag{2.6}$$

The steady-state behavior of AOB can also be visualized by allowing $\frac{\partial\rho}{\partial t} = 0$, and $\frac{\partial E_P}{\partial t} = 0$, so that the stationary field equation is given by [1,5]

$$\frac{\partial E_P}{\partial z} = -\chi(|E_P|^2)E_P, \tag{2.7}$$

where χ is the complex dielectric susceptibility, can be written as

$$\chi = \chi_a + i\chi_d. \tag{2.8}$$

The absorptive and the dispersive components of χ are defined as χ_a and χ_d, respectively. After some simplifications by calculating Eq.(2.1) in the steady-state and using definitions from Chapter 1 for the susceptibility, the expression of χ (similar to Eq.(1.28)) acquires the following form [1,5]

$$\chi = \alpha(1 - i\Delta)[1 + \Delta^2 + |E_P|^2/I_s]^{-1} = \alpha(1 - i\Delta)[1 + \Delta^2 + |x|^2]^{-1}, \tag{2.9}$$

where the unsaturated absorption coefficient $\alpha = \frac{2\pi\omega_P\mu_{21}^2 N}{\hbar c\gamma_{21}}$, the normalized atomic detuning parameter $\Delta = \frac{\omega_{21}-\omega_P}{\gamma_{21}} = \Delta_P/\gamma_{21}$, and the saturation intensity $I_s = \frac{\hbar^2\gamma_{21}\Gamma_{21}}{\mu_{21}^2}$. The parameter x is defined just after the Eq.(2.11) below.

In the mean-field theory one takes the limits [1,2]

$$\alpha L \to 0, \quad T \to 0, \quad \frac{\alpha L}{T} \to constant, \quad \delta_0/T = constant, \tag{2.10}$$

which means that electric field changes little in traversing through the medium, and it traverses the medium a large number of times to gain large nonlinearity. By integrating Eq.(2.7) over the medium length we get the steady-state equation for the AOB in a two-level atomic system as [1,2]

$$L\, E_P(L)\, \chi(|E_P(L)|^2) = -[E_P(L) - E_P(0)]$$
$$= -[E_P(L) - TE_P^I - RE_P(L)e^{-i\delta_0}]. \quad (2.11)$$

With the definitions of $y \equiv \frac{E_P^I}{\sqrt{I_s T}}$, $x \equiv \frac{E_P^T}{\sqrt{I_s T}}$, and rearranging the terms in Eq.(2.11) we get [1,2]

$$y = \frac{x(1 - Re^{-i\delta_0})}{T} + 2Cx\frac{1 - i\Delta}{1 + \Delta^2 + |x|^2}, \quad (2.12)$$

where $C = \frac{\alpha L}{2T}$ and L is the length of the nonlinear medium in the ring cavity [1,2]. When δ_0 is small (i.e., close to the cavity resonance mode) $e^{-i\delta_0} \sim 1 - i\delta_0$ and therefore we get

$$y = x[1 + 2C\bar{\chi}_a] + ix[\theta_0 + 2C\bar{\chi}_d], \quad (2.13)$$

where $\theta_0 = R\delta_0/T$, $\bar{\chi}_a = \chi_a/\alpha$ and $\bar{\chi}_d = \chi_d/\alpha$.

2.1.1 *Split in transmission spectrum of the optical cavity*

We consider a situation in which the atomic transition frequency is in resonance with the cavity mode frequency, e.g., $\omega_{21} = \omega_{cav}$. Thus we have $\Delta_P = \Delta\gamma_{21} = \kappa\theta_0$, where κ represents cavity damping coefficient defined in terms of cavity transmission coefficient and cavity round-trip time [1]. We can rewrite Eq.(2.13) using these definitions in the following way [81-83]

$$x = y\left[\frac{\kappa(\gamma_{21} + i\Delta_P)}{(\kappa + i\Delta_P)(\gamma_{21} + i\Delta_P) + \frac{g_{21}^2 N_a}{(1 + \gamma_{21}^2 |x|^2/(\gamma_{21}^2 + \Delta_P^2))}}\right]. \quad (2.14)$$

In Eq.(2.14) above, $N_a = NV$ represents total number of atoms. The transmission function of the cavity can be thus written as [81-83]

$$\left|\frac{x}{y}\right|^2 = \left|\frac{F}{i\Delta_P + \Delta_1} + \frac{G}{i\Delta_P + \Delta_2}\right|^2, \quad (2.15)$$

in which

$$F = \kappa\frac{\gamma_{21} + \Delta_1}{\Delta_1 - \Delta_2}, \quad G = \kappa\frac{\gamma_{21} + \Delta_2}{\Delta_2 - \Delta_1},$$

$$\Delta_{1,2} = -\frac{\kappa + \gamma_{21}}{2} \pm i\left[-\left(\frac{\kappa - \gamma_{21}}{2}\right)^2 + \frac{g_{21}^2 N_a}{(1 + \gamma_{21}^2 |x|^2/(\gamma_{21}^2 + \Delta_P^2))}\right]^{1/2}. \quad (2.16)$$

This expression clearly brings out the presence of two normal modes in the weak intensity regime and strong coupling condition or good cavity limit, i.e., $g_{12} >> \kappa, \gamma_{21}$ [81-83]. In the limiting case of extremely low excitation, $|x|^2 << 1$, the atom-cavity system behaves like two coupled harmonic oscillators and $\Delta_{1,2}$ are the eigenvalues of the linearized Maxwell-Bloch equations representing a mode splitting of this combined system. In the absence of atom-cavity detuning the separation of these modes from atomic resonance is in the order of $\Omega_{vac} = g_{21}\sqrt{N_a}$ on each side, where Ω_{vac} is called vacuum-Rabi frequency of the system [81-83]. When a weak-field limit is considered, the imaginary part of Δ_{12} represents the coupling between the cavity and atoms. With intensity going up inside the cavity, the coupling between field and atoms reduce down. This makes atomic response more like single atoms rather than a collective response up to the level of saturation. When high excitation limit is reached, i.e., ($|x|^2 >> 1$), the atoms are saturated and hence they do not have any significant role in the cavity transmission profile. In such situation the imaginary part of Δ_{12} goes real and the transmission profile is thus governed by the real part of Δ_{12} [81-83]. The transmission profile of cavity under this limiting condition looks like that of an empty cavity transmission profile. Also, the vacuum Rabi peaks evolves from a doublet to a singlet because atoms are not taking part in the cavity field evolution.

The physical understanding of these behaviors are outlined in Refs.[81-84] using semiclassical picture. Semiclassically the atoms act as a medium providing refractive index and thus introducing a phase shift in the electric field. Under an appropriate laser frequency detuning, the frequency shift imposed by the cavity and the atomic phase shift will counterbalance each other. The cavity transmission becomes intensified because of the constructive interference. On the other hand, for weak input intensity and low atom-cavity detuning this zero-phase shift condition can occur at three different frequencies. One is on resonance which locates in the center, and other two transmission peaks on either side to the center peak are the peaks of vacuum Rabi splitting [81-84]. The intensity of transmission is considerably reduced at central peak due to a large atomic absorption.

The positions of peaks under this zero-phase condition can be calculated

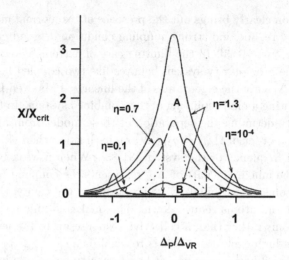

Fig. 2.2. Evolutions of the two vacuum Rabi peaks depending on the input intensity $\eta = Y/Y_{crit}$. The solid curves are solutions of the state equation (2.13) and the dashed curve is the zero-phase condition (2.17)(adopted from [83] with permission).

using Eq.(2.13) under zero atom-cavity detuning, which are given by [81-83]

$$\Omega_{vac}^{G} = \pm\gamma_{21}\sqrt{\frac{8X}{3[[2\exp(\frac{2\gamma_{21}^{2}X}{3g_{21}^{2}N_{a}}) - 1]^{2} - 1]} - 1}, \qquad (2.17)$$

for the Gaussian standing-wave model and

$$\Omega_{vac}^{P} = \pm\gamma_{21}\sqrt{\frac{g_{21}^{2}N_{a}}{\gamma_{21}^{2}} - 1 - X}, \qquad (2.18)$$

for the traveling plane-wave model. The above two expressions are justified provided Ω_{vac}^{G} is real and for that the intensity $X = |x|^{2}$ should be small enough. The cavity transmission peaks are located at slightly different positions from what are predicted from zero-phase model. The evolutions of the vacuum Rabi peaks are shown in Fig.2.2, where the zero-phase condition for zero atom-cavity detuning has been included. The plots in this figure are for the Gaussian standing-wave model and the x-axis represents the normalized detuning Δ/Δ_{VR} ($\Delta_{VR} = \Omega_{vac}$). Note that the output intensity is normalized to X_{crit}. Here X_{crit} means the smallest intensity required for atoms to get saturated under the on resonance condition. At this value of intensity the system jumps from the upper branch to the lower branch of the bistability hysteresis loop under the on-resonance condition, alternatively it can be called as the lower threshold of the optical bistability.

The input intensity corresponding to the output intensity X_{crit} is given by Y_{crit} [81-83].

When input intensity is small, e.g., $\eta = Y/Y_{crit} = 10^{-4}$, the cavity transmission spectrum basically consists of two vacuum Rabi peaks. By increasing the intensity in the cavity, there is saturation effect for atoms, consequently the peaks shift their positions and started having distorted profiles. The behavior of system looks like an anharmonic oscillator type and peaks leans towards the center. The peaks shows merger under large cavity input intensity. Interestingly, when input intensity is in the intermediate range then from Eq.(2.13) one gets three output intensity values in a specified range of laser detuning. For $\eta = 0.7$, instability has been observed as shown in the dotted curves. The phenomenon of instability will be dealt extensively in Chapter 5. The two vertical dashed arrows in the Fig.2 show that when frequency scan is made from lower to higher value the system goes from the upper branch to the lower branch of the hysteresis curve and back [81-83]. As intensity goes higher ($\eta = 1.3$) the regular one-peaked spectrum A and a cave-shaped one B are observed (see Fig.2.2). Here, in this figure, the parametric conditions are $C = 70$, $(\kappa, \gamma_{21}, g_{21}) = (3.0, 3.1, 2.5) \times 2\pi \times 10^6$ rad/s, $X_{crit} = 390$, and $Y_{crit} = 2500$. Clearly, the system evolves from two coupled harmonic oscillators like situation at low intensity to highly deformed anharmonic oscillators at higher intensity as shown in Fig.2.2. More details of this anharmonicity can be found in Ref.[81]. When the atom-cavity detuning has a finite non-zero value then coupling between cavity field and atoms reduces. This makes alteration in the position and height of the peaks [81-84]. The details of experiments and obtained results are matching very well with the theoretical model discussed above are available in Ref.[83]. When Finesse of the cavity is low, the system of two-level atoms inside such cavity exhibits the phenomenon of AOB, which will be described in the following using the basic formulation given above.

2.2 Atomic Optical Bistability

Atomic systems with nonlinearities in absorption (saturation) and/or dispersion (in refractive index) can have more than one stable output state for a certain input state. When two stable output intensity states exist for a given input state then such optical system exhibits optical bistability. If there are more than two such stable output intensity states then the system exhibits optical multistability [1-4]. In such bistable and multistable systems, these output states can be used to implement all-optical switching and exhibit interesting dynamic behaviors.

2.2.1 Atomic optical bistability: the mean field theory

The theoretical prediction of optical bistability was first given by Szóke et al [85] and the first experimental observation was made by Gibbs et al [6] using a sodium-filled FP interferometer. The description of this phenomenon was discussed using a nonlinear optical medium inside a FP resonator. The FP resonator was discussed in Chapter 1 (section 1.4). A ring cavity or traveling-wave cavity behaves like a FP resonators, so the optical ring cavity has been explicitly used above in section 2.1 to describe the interaction between a collection of atoms and an electromagnetic field mode of the cavity. We concentrate on Eq.(2.13), which gives a relationship between the cavity input field amplitude (y) and intracavity field amplitude (x), both normalized to the saturation field amplitude for the atoms. The relationship between the intracavity intensity ($X = |x|^2$) and input intensity ($Y = |y|^2$) is given by [1,2]

$$
\begin{aligned}
Y &= X[(1 + 2C\bar{\chi}_a)^2 + (\theta_0 + 2C\bar{\chi}_d)^2] \\
&= X[(1 + \frac{2C}{1 + \Delta^2 + X})^2 + (\theta_0 - \frac{2C\Delta}{1 + \Delta^2 + X})^2],
\end{aligned}
\tag{2.19}
$$

for the homogeneously broadened system. This Eq.(2.19) is exactly similar to the so called state equation of optical bistability using mean field model of optical bistability [1,2]. Hence Eq.(2.10) gives the so called 'mean field limit', which assumes that quantities like wave vector of field and absorption of atoms are spatially invariant going through one round trip in the cavity. Equation (2.19) was obtained for a homogeneously-broadened system [1] and another explicit expression was also given for Lorentzian inhomogeneously-broadened system [1]. The above expression (Eq.(2.19)), which connects incident intensity to transmitted (intracavity) intensity, could also be given phenomenologically [6,85]. The description provided here is an analytical one using first principle approach by solving Maxwell-Bloch equations along with the field boundary conditions under the mean field approximation as described above in Eqs.(2.4) and (2.10), respectively. What follows next is the discussion on the interpretation of mean field approximation [1]. The first one in this approximation stated as $\alpha L \to 0$

(which means $\alpha \to 0$), is showing a weak coupling limit between the atoms and the field traversing the cavity. Under the specific conditions, $\alpha L \to 0$ but T finite, we get $C = \alpha L/(2T) = 0$. This gives the empty cavity result, i.e., $Y = X(1 + \theta_0^2)$. But when $T \to 0$, then parameter C becomes arbitrary and provides the nonlinear terms in Eq.(2.19), which gives rise to many noteworthy phenomena [1]. The meaning of $T \to 0$ indicates infinite life-time of photons inside the cavity. Finally, the limit $\delta_0 \to 0$ (or $\theta_0 = $ finite) implies that the magnitude of cavity detuning is lesser than the FSR of the cavity. However, it should be almost equal to the magnitude of the cavity linewidth cT/L_T. This means that such a system will give desired results provided the cavity mode is almost on resonance with the incident input field to the cavity [1].

One point to be noticed about the mean field approximation is even in the limit $\alpha L \to 0$, Eq.(2.19) does not imply strictly the weak coupling theory. The limit mentioned in Eq.(2.10) is a bit unusual one in the sense that field inside the cavity becomes infinite but the normalized variables X and Y remain finite. In fact the internal cavity field is quite different from the incident field as it has been produced due to cooperative reactions of the atoms inside the cavity [1].

2.3 Absorptive Atomic Optical Bistability

2.3.1 *Simple model of absorptive atomic optical bistability*

Some simple models were developed to explain the absorptive and dispersive optical bistabilities. The model given by Szöke et al [85], for the absorptive optical bistability is discussed next. The basic philosophy behind this model is the inequality condition obtained from the saturation equation of a homogeneously-broadened two-level system and the boundary conditions of field inside an optical cavity. In order to obtain bistability the inequality needs to be satisfied. This inequality condition has been derived [2] here and given at the end of this subsection. The simplified diagram of a plane FP cavity with spacing between mirrors to be L and the reflectivities of the mirrors are assumed to be R is shown in Fig.2.3. The FP cavity is containing a saturable absorber. The laser (having frequency ω_P) enters the FP cavity (having frequency ω_{FP}) and travels back and forth in the cavity and in this model effects due to the standing-wave as well as intensity-dependent refractive index are ignored just for the sake of simplicity [2].

The boundary conditions in this case are

$$E_T = \sqrt{T}E_A(L),$$
$$E_A(0) = \sqrt{T}E_I + Re^{i\phi}e^{-\frac{\alpha}{2}(2L)}E_A(0), \tag{2.20}$$

in which the cavity-laser phase detuning and absorption coefficient are given by the parameter ϕ and α, respectively. Here, E_I, E_R, E_A, E_B, and E_T represent slowly-varying complex field amplitudes of incident, reflected, forward, backward, and transmitted electric fields, respectively [2]. For the purely absorptive AOB it is convenient to use $\omega_P = \omega_{FP}$ and also select $e^{i\phi} = 1$. From Eq.(2.20) one gets [2,3]

$$\frac{E_A(0)}{E_I} = \frac{\sqrt{T}}{1 - Re^{-\alpha L}}. \tag{2.21}$$

To simplify the analysis absorption can be taken as very small, i.e., $\alpha L <<$ 1 but the empty cavity Finesse is sufficiently high so that the quantity $k = \frac{R\alpha L}{(1-R)} >> 1$ (where L is the length of atomic medium in the FP device, which equals cavity length because atomic medium fills the entire cavity length [1,2]), that provides

$$\frac{E_A(0)}{E_I} = \frac{\sqrt{T}}{1 - R(1 - \alpha L)} = \frac{1}{\sqrt{T}(1 + k)}. \tag{2.22}$$

The transmitted field E_T can be written as

$$E_T = \sqrt{T}E_A(L) \simeq \sqrt{T}E_A(0), \tag{2.23}$$

which eventually leads to a relationship between the transmitted and incident intensities as

$$\frac{I_T}{I_I} = \frac{|E_T|^2}{|E_I|^2} \simeq \frac{1}{(1 + k)^2}. \tag{2.24}$$

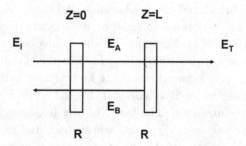

Fig. 2.3   Electric fields in a FP resonator.

At large intensities both α and k are intensity dependent. For a two-level system with upper-level population N_2 and lower-level population N_1, without the cavity, the rate equation reads

$$\dot{N_2} = -\frac{N_2}{T_1} + \frac{\alpha I_A}{\hbar \omega_\rho}, \tag{2.25}$$

in which T_1 is the homogeneous relaxation time (Γ_{21}^{-1}). The quantity $\frac{\alpha I_A}{\hbar \omega_\rho} = \frac{energy/(volume \times time)}{energy\ per\ atom}$ is the number of atoms excited per unit volume per unit time. Under the steady-state condition, $\dot{N_2} = 0$ so that $N_2 = \frac{\alpha I_A T_1}{\hbar \omega_\rho}$. The absorption coefficient α can be obtained from the equation of stimulated emission [2],

$$\alpha = \frac{N_2 - N_1}{N_2 + N_1} \alpha_0 = \frac{\alpha_0}{1 + \frac{I_A}{I_S}}, \tag{2.26}$$

in which α_0 is the unsaturated absorption coefficient and I_S is the saturation intensity:

$$I_S = \frac{(N_1 + N_2)\hbar\omega_\rho}{2T_1\alpha_0}. \tag{2.27}$$

At this stage k can be written as

$$k = \frac{k_0}{1 + \frac{I_A}{I_S}}, \quad k_0 = \frac{R\alpha_0 L}{1 - R}, \tag{2.28}$$

and Eq.(2.24) can be expressed as follows [2]

$$Y = X[(1 + \frac{k_0}{1 + X}]^2. \tag{2.29}$$

Such absorptive AOB equation can also be derived using the cavity transmission function at the limit of dominating absorption system [3]. In the Eq.(2.29), $Y = I_I/TI_S$ and $X = I_T/TI_S$ are defined as the normalized input (incident) and output (transmitted) intensities, respectively (consistent with the definition of x and y defined just above Eq.(2.12)). The condition of bistability is $\frac{dI_I}{dI_T} < 0$ or $\frac{dY}{dX} < 0$. By differentiating Eq.(2.29), and setting $\frac{dY}{dX} = 0$, the following values of X are obtained [2]:

$$X = (1/2)(k_0 - 2 \pm \sqrt{[k_0(k_0 - 8)]}). \tag{2.30}$$

In order to have zero slope for physically meaningful value of X, the value of k_0 must equal or greater than 8. Hence, the condition for observing absorptive AOB [2] is

$$k_0 \equiv \frac{\alpha_0 LR}{1 - R} > 8. \tag{2.31}$$

Equivalently, as $R \simeq 1$ and $R + T = 1$,

$$C \equiv \frac{\alpha_0 L}{2T} > 4. \tag{2.32}$$

where C is the cooperativity parameter, defined earlier and L is the length of nonlinear absorbing material in the FP cavity which is also the total length of cavity in this case.

2.3.2   *Mean field theory of absorptive atomic optical bistability*

The mean field theory of AOB was given by Bonifacio and Lugiato [86] in which the reaction field produced by atoms due to cooperative effect acts against the incident field. The absorptive AOB can be described by Eq.(2.19) considering resonant conditions $\Delta = \theta_0 = 0$. Equation (2.19) can be recast in terms of field amplitudes (x and y) under this resonant condition as

$$y = x + \frac{2Cx}{1 + x^2}. \qquad (2.33)$$

The second term is the resonant reaction field arising from the atomic co-operation. The yardstick of atomic cooperativity is through the parameter C. It is easy to see that under the limit when input field x is quite large then atoms get saturated in absorption and the medium provides transparency leading to the situation $y = x$, i.e., empty cavity situation. Under this condition there is no cooperation among atoms. Also, the atomic correlations are negligibly small and each of the atoms acts as an individual atom to interact with the electromagnetic field. In the other limit when x is small the relationship between x and y goes as $y = (1 + 2C)x$. This is a linear relationship between the two fields as atoms are unsaturated. When C becomes large the atom-atom correlations increase, giving rise to the absorptive OB feature from the system [1,2,80]. In Fig.2.4, y is plotted as a function of x for different values of cooperativity parameter C. Curves I, II, and III are for $C = 3, 4$, and 6, respectively. When $C < 4$, y behaves like an unchanging function of x and hence bistability is not apparent. This curve resembles to the characteristic curve of a transistor (slope at some part of the curve: $dx/dy > 1$) and hence the system behaves like an optical transistor. When $C = 4$, the curve II shows a point of inflection with a horizontal tangent and this is at the criticality condition to observe optical bistability. Clearly, when $C > 4$ (Curve III), there is bistability in this curve. In fact this curve shows a maximum and a minimum situated at $(x_{max}, y_{max}) \sim (1, C)$ and $(x_{min}, y_{min}) \sim (\sqrt{2C}, \sqrt{8C})$, respectively. So between y_{min} and y_{max}, there are three stationary solutions (x_a, x_b, x_c). The solution x_b lies on the negative slope part of the curve III and hence unstable [1,80]. Only two solutions x_a and x_c are stable. When we exchange the axes x and y, a typical hysteresis curve between the transmitted and incident fields is obtained, representing the absorptive AOB under resonant condition. The states x_a and x_c are termed as 'cooperative stationary

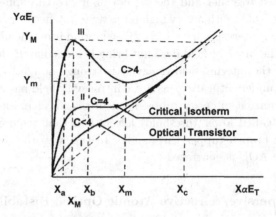

Fig. 2.4   Plots of the mean-field state equation for pure absorptive bistability with $\theta = 0$, for different values of cooperativity parameter and $X_M = x_{max}$, $X_m = x_{min}$ etc (adopted from Fig.8 of [1] with permission).

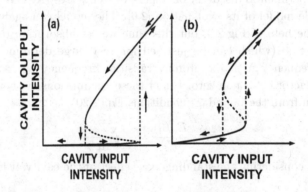

Fig. 2.5   Hysteresis loop of (a) a typical absorptive AOB and (b) a dispersive AOB.

state' and 'one-atom stationary state', respectively, as atomic cooperation dominates in the former [1,2]. The typical absorptive AOB is plotted in Fig.2.5(a). The physical mechanism involved in absorptive optical bistability under the resonant condition $\Delta = \theta_0 = 0$ is summarized as follows. In the lower branch of AOB hysteresis curve the transmission of intensity is small because atomic medium absorbs the incident field as it is on resonance. Further increase in the input intensity causes atoms to saturate. Hence the internal cavity field increases rapidly due to multiple reflections from cavity mirrors. At certain incident intensity all the atomic absorber are bleached together and the system jumps to the upper branch (called upper threshold) with a very high intracavity intensity or high transmission part of the hysteresis curve [1,2]. When system is on the upper branch then reducing the incident field does not bring it immediately down to the lower branch as the internal field of cavity is strong enough to keep the atomic absorbers under saturation (as if the intracavity field has certain memory). The transmitted intensity drops down to the lower branch at an incident intensity (called lower threshold) lower than that required to move it to the upper branch (upper threshold). In this way the hysteresis cycle in absorptive AOB is generated.

2.4 Dispersive/refractive Atomic Optical Bistability

2.4.1 *Simple model of dispersive/refractive atomic optical bistability*

The experimental discovery of dispersive (refractive) AOB in sodium led to a simple model of its explanation [2,6]. The model can again be described with the help of Fig.2.3, but this time we set absorption to be very low, i.e., $\alpha \simeq 0$ (which can be justified for very large detuning of the cavity field frequency with the atomic transition frequency ($\Delta >> 0$)) but the phase factor $e^{i\phi}$ is non-zero. Under these assumptions, we get the following relation from the boundary conditions Eq.(2.20)

$$\frac{E_A(0)}{E_I} = \frac{\sqrt{T}}{1 - Re^{i\phi}}. \tag{2.34}$$

By not considering the standing-wave effect of the cavity, it is easy to show that [2]

$$E_T = \sqrt{T}e^{i\phi/2}E_A(0), \quad E_I = \frac{(1 - Re^{i\phi})E_T}{Te^{i\phi/2}}, \tag{2.35}$$

which leads to the following relationship between input (I_I) and output/transmitted (I_T) intensities

$$\frac{I_I}{I_T} = \frac{|1 - Re^{i\phi}|^2}{T^2}. \tag{2.36}$$

In the close proximity of the transmission peak, assuming ϕ to be small and $T << 1$, the relationship between input and output intensities goes as [2]

$$\frac{I_I}{I_T} \simeq (1 + R\phi^2/T^2), \quad or \quad Y \simeq X(1 + R\phi^2/T^2). \tag{2.37}$$

The phase ϕ could be intensity dependent (e.g., in the case of Kerr non-linearity): $\phi = \phi_0 + \phi_2 I_T$ (see Eq.(2.47) below or Eq.(1.49)). The above equation (2.37) can also be obtained from the general equation of cavity transmission formula (Eq.(2.46) below or Eq.(1.48)) under the conditions of $\alpha = 0$ and for small δ (which is ϕ here) [3]. Note that the I_T and intracavity field intensity I_{cav} are related with each other. The Kerr nonlinear index n_2 is defined through the equation $n \backsim n_0 + n_2 I_T$, in which n is the total refractive index and n_0 is linear refractive index. The dependence of the phase ϕ_2 on such nonlinear index n_2 and hence the intensity-dependent total phase becomes responsible for the dispersive (refractive) optical bistability [1-3].

2.4.2 *Mean field theory of dispersive/refractive atomic optical bistability*

Under the resonant condition of the optical field with atomic transition frequency (i.e., atomic detuning: $\Delta = 0$), a pure absorptive AOB is observed as discussed in section 2.3. On the other hand if the atomic detuning is very large pure dispersive (refractive) AOB could be observed as the absorptive part of the susceptibility (Eq.(2.8)) becomes vanishingly small. The bistability state equation (Eq.(2.19)) under this condition becomes [1,86]

$$Y = X[1 + (\theta_0 - \frac{2C\Delta}{1 + \Delta^2 + X})^2].\qquad(2.38)$$

If the conditions $\Delta^2 \gg 1$, $\Delta\theta_0 \gg 1$, and $\Delta \gg \theta_0$ are imposed [1] assuming homogeneously-broadened two-level system satisfying

$$2C/(\Delta\theta_0) - 1 \ll 1,\qquad(2.39)$$

then expanding: $(1 + \Delta^2 + X)^{-1} = \Delta^{-2}(1 + \frac{1+X}{\Delta^2})^{-1} \simeq (\frac{1}{\Delta^2} - \frac{X}{\Delta^4})$ in Eq.(2.38), it simplifies to [1]

$$Y = X[1 + (\theta_0 - 2C/\Delta + 2CX/\Delta^3)^2].\qquad(2.40)$$

The above equation displays bistable behavior under the condition $(2C/\Delta) - \theta_0 > \sqrt{3}$. It is ensured by the condition of Eq.(2.39) that the atoms do not move out from the ground state meaning no significant absorption throughout the entire bistable regime. The cubic model of pure dispersive (refractive) AOB obeys the following output-input intensity relationship [6]

$$Y = X[1 + (P - QX)^2],\qquad(2.41)$$

which exhibits bistability under the condition $P > \sqrt{3}$. The cubic model explains optical bistability in such Kerr-like medium and Eq.(2.40) is quite similar to this model using mean field theory. The typical dispersive (refractive) AOB is plotted in Fig.2.5(b).

The physical mechanism responsible for dispersive (refractive) AOB [6] is not the same as given above for the absorptive AOB. Since the frequency of the empty cavity ω_{cav} is far detuned from the ingoing field frequency ω_P so the transmission is low on the lower branch of the hysteresis curve. By enhancing the cavity input field intensity, the nonlinear refractive index of the atomic medium (which increases as $|E_P|^2$) causes effective optical length of the cavity to change and brings cavity towards resonance with the ingoing field frequency provided the atomic and cavity detunings are having the same sign [1]. The growing of the intracavity field further enhances the tendency of the cavity to move towards frequency of the ingoing optical field until resonance condition is satisfied and field jumps from the lower branch to upper branch of the hysteresis loop. When the system is on the upper branch of the hysteresis loop, reducing down the incident field does not bring cavity out of the resonance with the cavity field because the intracavity field is still strong enough to maintain resonance. It requires further decreasing of the incident field to switch down to the lower branch which surpasses the upper switching threshold value and hence a hysteresis cycle is formed [1]. At this stage it necessitates to ask the question about the relative advantages of the absorptive versus dispersive AOB. To observe dispersive (refractive) AOB , the saturation of atomic medium is not necessary. There are practical problems in achieving resonance conditions in steady-state between atoms and the laser field because each absorptive and dispersive AOB has its own properties, depending on the type of systems under different conditions.

2.5 Mixed Absorptive-dispersive Atomic Optical Bistability

A general case for Eq.(2.19) is when Δ and θ_0 are both non zero. When the product of quantities $\Delta\theta_0 < 0$, it is not quite simple to observe optical bistability. However, when $\Delta\theta_0 > 0$, then situation is symmetrical with respect to simultaneous sign change in the quantities Δ and θ_0. But we assume both $\Delta, \theta_0 \geq 0$. The expression Y as a function of X given in Eq.(2.19) has an inflection point (X_{IF}) at [1]

$$X_{IF} = \frac{2C - \Delta\theta_0 + 1}{C + \Delta\theta_0 - 1}(\Delta^2 + 1). \tag{2.42}$$

To observe bistability the necessary conditions are

$$X_{IF} > 0, \quad \frac{dY}{dX}\Big|_{at \ X=X_{IF}} < 0. \tag{2.43}$$

The first condition ensures that the point of inflection lies within the region $X > 0$. From the second condition it is possible to determine the parameters for which the curve Y when plotted as a function of X has a minimum and maximum values [1]. When $\Delta\theta_0 > 0$, the first condition in Eq.(2.43) provides

$$2C > \Delta\theta_0 - 1. \tag{2.44}$$

Using the other condition, one gets

$$(2C - \Delta\theta_0 + 1)^2(C + 4\Delta\theta_0 - 4) > 27C(\Delta + \theta_0)^2. \tag{2.45}$$

By carefully analyzing these conditions [1] it is easy to infer that (a) it is not possible to observe bistability for $C < 4$. (b) When certain value of C is selected such that $C > 4$, then the biggest hysteresis curve is obtained for $\Delta = \theta_0 = 0$ (see discussion in section 2.3). Under this condition optical bistability exists only in a limited domain around origin in the (Δ, θ_0) plane. (c) When $C > 4$ is kept as fixed and both Δ, θ_0 increase from 0, while keeping Δ/θ_0 fixed, the hysteresis curve of X versus Y moves towards left and its width etc decreases and eventually it vanishes. (d) When C and Δ kept as fixed and C satisfies the condition of Eq.(2.45) for $\theta_0 = 0$, then increasing θ_0, the size of hysteresis cycle increases to a maximum and then reduces and goes to zero when $\theta_0 < (2C + 1)/\Delta$ [1].

At this stage one can draw a conclusion for the homogeneously-broadened two-level systems, i.e., if such a system does not show absorptive bistability at $\theta_0 = 0$, then it will also not exhibit dispersive bistability for the general non-zero values of Δ and θ_0. However, this condition is not applicable for the inhomogeneously-broadened systems which depend on polarization decay time (T_2^*). For the situation $(\gamma_{21}T_2^*)^{-1} >> 1$, there is no bistability for the values $\Delta = 0$, $\theta_0 = 0$, but for the large values of these quantities the bistability is observable, meaning no absorptive bistability but only dispersive bistability can be displayed by the system [1]. In the homogeneously-broadened system the size of hysteresis cycle is largest when $\Delta = \theta_0 = 0$, which is not applicable for the inhomogeneously-broadened system. If we set $\Delta, \theta_0 \neq 0$, the upper threshold value comes down considerably, i.e., the switching occurs at low level of incident intensity. Thus dispersive optical bistability provides the advantage of having a low upper switching threshold and prevents excessive excitations of the atomic medium [1].

The definitions attributed to the absorptive and dispersive (refractive) AOB
are according to whether their generations are originated from the absorptive and dispersive parts of the susceptibility [1]. This should not be interpreted as if there is prominent absorption in absorptive AOB and absence
of absorption in dispersive AOB. The energy going in to the cavity is partly
transmitted, partly reflected and partly absorbed by the atomic vapor inside
the cavity. The energy absorbed is either diffused in fluorescence emission
or could be dissipated in the atomic vapor. In the absorptive bistability
case most of the light is transmitted when the system is in the upper state
and there is large absorption by the atomic sample when it is in the lower
state along with reflection from cavity mirrors. There are situations in pure
dispersive (refractive) AOB where population of the upper level is appreciable. However, when conditions mentioned above Eq.(2.39) are satisfied,
the absorption is completely absent in such dispersive AOB [1].

The hysteresis cycle of AOB is normally observed by varying input intensity
and measuring the output intensity, keeping other experimental parameters,
e.g., C, Δ, and θ_0 unchanged. It is also possible to obtain the hysteresis
curve by selecting the input intensity to a fixed value and varying other
parameters C, Δ, and θ_0 etc. One can vary Δ, and θ_0 simultaneously by
adiabatically changing the incident field frequency [1].

Alternatively, the phenomenon of AOB can also be explained using
Eqs.(1.48) and (1.49), which we reproduce in the following (with a slight
change in exponential factor for the ring cavity) as

$$I_T = \frac{T^2 e^{-\alpha L} I_I}{(1 - Re^{-\alpha L})^2 + 4Re^{-\alpha L} \sin^2(\delta/2)}. \tag{2.46}$$

In above expression the coefficient α represents extinction of light per unit
length (related to the absorptive part of the susceptibility χ_a) and the total
absorption per round-trip is $e^{-\alpha L}$. The round-trip phase is given by

$$\delta = \frac{\omega_P l}{c} + \frac{(n_0 - 1)\omega_P L}{c} + \frac{(n_2 I_{cav})\omega_P L}{c} - 2\pi m, \tag{2.47}$$

where l is the total length of the cavity, L is length of atomic medium placed
inside the cavity, usually in a glass cell, I_{cav} is the intensity of the field
inside the cavity, and refractive index $n = n_0 + n_2 I_{cav}$, composed of linear
part (n_0) and nonlinear part (n_2) multiplied by the field intensity inside
the cavity. This part is related to the dispersive part of the susceptibility.

Note that Eq.(2.46) above is also referred to as the response function of the optical ring cavity with an atomic vapor cell of length L placed in one arm of the cavity. This is a general expression. The absorptive and dispersive limits of AOB can both be obtained from this response function, under different limits [3].

2.6 Experimental Demonstrations of Two-level Atomic Optical Bistability

Since the theoretical explanation of absorptive AOB came out very well, attempts were made soon after for its experimental observation. Initially no AOB could be observed but nonlinear transmission and pulse re-shaping were noticed [2]. The first experimental observation of passive AOB was made in sodium atomic vapor contained in a FP cavity [2,6]. In the experiment both the absorptive and dispersive AOB were discovered and the observed dispersive AOB could be well understood. It also proved that the inhomogeneous-broadening and non-uniform transverse profile of the field do not disallow observation of AOB. These experiments also demonstrated optical logic functions, e.g., optical transistor, clipping, and limiting action. To observe the dispersive (refractive) AOB, a custom design was made for a Na vapor chamber in which an inert gas (Argon) was flowing. This gas was used for two purposes, viz., providing homogeneous broadening to Na and preventing Na vapor to reach FP reflecting mirrors [2]. The setup is shown in Fig.2.6. The laser used here was a cw dye laser pumped by a 3 Watt argon laser. The laser was frequency stabilized with the help of an external

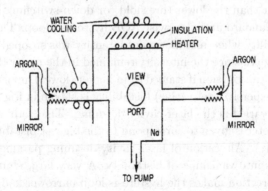

Fig. 2.6 Sodium optical bistability apparatus (adopted from Fig.3.2-1 of [2] with permission).

empty FP cavity. To calibrate the laser frequency, a saturation absorption spectroscopy setup was employed using a sodium vapor cell. Typical power reaching the FP cavity input was of the order of 10-13 mW. The laser with a beam diameter of 1.65 mm was used. This beam diameter was large enough so that the diffraction losses could be neglected for the multiple passes through the FP cavity. The mirrors used in the FP cavity were flat with dielectric coating having reflectivity of 90%. One of the mirrors in the FP resonator was mounted on a PZT for FP cavity length scanning. Water cooling of the sodium atomic oven was provided so that thermal effect in changing cavity length could be avoided. Despite all these cares and precautions the system was marginally stable for certain experimental observation such as ac gain etc [2].

To observe AOB features, an acousto-optic modulator (AOM) was used so that the input intensity could be modulated in a triangular wave. For fast-switching the square-step modulation was used. The input and output (transmitted) intensities were measured by phototubes and displayed on an x-y oscilloscope. Initial measurements were made without FP mirrors. The curve I_T vs I_I shows saturation and it is a straight line inclined at an angle 45^0 approximately with respect to the x-axis. The saturation effect in the absence of Argon is due to the optical pumping whose relaxation time is quite large compared to the radiative life time of a two-level system [2]. Addition of argon buffer gas at this stage reduces the optical pumping and hence the saturation effect of homogeneously-broadened line. When FP mirrors were kept in place and without using argon gas, AOB could be observed. The upper threshold (or up-switching intensity) was found to be larger than the lower threshold (or down-switching intensity), which formed a standard anti-clockwise AOB hysteresis loop. The system showed true bistability when modulation of intensity was stopped in the middle of bistable region, since the intensity remained in the upper (lower) state with high (low) transmission if started from higher (lower) intensity level [2]. The observed dispersive (refractive) bistability is shown in Fig.2.7, in which the parameter varied is the laser-cavity detuning. The 'zero' assigned value of this parameter is given to correspond to the biggest bistable hysteresis loop (erroneously). Alteration of laser-cavity detuning parameter in one direction results into vanishing of bistability. A very large shifts of detuning in opposite direction makes the hysteresis loop narrower and upper threshold reduces. By further increase in cavity detuning in the same direction causes bistability to vanish altogether. These results are the desired ones for the dispersive bistability. On the other hand the pure absorptive AOB occurs because of the saturation effect and it is obvious that the hysteresis cycle of bistability will changes its shape and width symmetrically with respect to the change in detuning [2]. When these experiments were carried out, only absorptive AOB was well understood. However, the asymmetry observed in this figure with respect to the cavity detuning clearly shows the presence of the dispersive characteristics of the bistability. This was further confirmed by observing the transmission peak (which was symmetrical) of the FP cavity on the oscilloscope with respect to FP cavity tuning. The transmission peaks from the empty cavity was found to be symmetrical in shape and equally displaced when laser was detuned to ±360 MHz. Addition of sodium atomic vapor in the FP cavity caused asymmetry in the shape of peaks and their unequal displacements with laser detuning. This could be explained from the intensity-dependent nonlinear refractive index (Kerr nonlinear index) of sodium atomic vapor [2].

After the initial successful experimental demonstration of dispersive AOB, several other experiments were reported over the years for investigating the AOB in two-level atomic systems by various groups [1,2]. The pure absorptive AOB in a collection of radiatively-broadened two-state atoms was systematically investigated in atomic beams interacting with a single mode of cavity. This experimental system was having well defined qualities so that absolute comparisons with the theory could be made [87]. In these experiments non-resonant conditions for both cavity and atoms were also

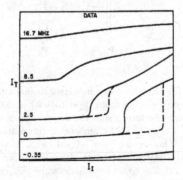

Fig. 2.7 Sodium transmission for various laser-cavity detunings in megahertz, where zero detuning has been arbitrarily (and erroneously) assigned to the largest bistability loop (adopted from Fig.3.2-4 of [2] with permission).

considered so that the absorptive and dispersive contributions of susceptibility were present simultaneously allowing detail comparisons with the
theory including the spatial propagation effects [88]. A brief discussion is
provided in the following for these experiments.

The later experiments on AOB in two-level atomic systems were carried out using atomic beams to eliminate the first-order Doppler effect (in
a best possible manner), which hampers the observation of AOB in a two-
level atomic system. In the initial experiments [2,6] a special system was
designed in which a buffer gas provided reduction of Doppler broadening.
The experimental system (Fig.2.8) contained a group of ten well-collimated
optically-prepumped atomic beams of sodium atoms crossing a high-finesse
cavity axis at an angle of 90^0 [5,87,88]. Initially the sodium atoms were

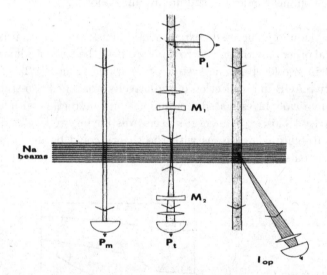

Fig. 2.8   Experimental setup showing three laser beams intersecting ten well-collimated
atomic beams of sodium (moving from right to left) at 90^0. The first laser on the far
right optically prepumps the atoms into the $3S_{1/2}, F = 2, m_F = 2$ state of the D_2 line,
with the fluorescence (grainy shade) from one of the center beams (I_{op}) collected on a
photomultiplier tube. The second laser (signal beam in the middle) is mode-matched
to the cavity formed by mirrors M1 and M2. The input power (P_i) is measured by
splitting a constant fraction onto a photodiode while the output power (P_t) transmitted
through the cavity is detected by a photomultiplier tube. Further downstream (far left)
the small-signal absorption is measured with the monitor beam (P_m). The transverse
dimensions of the laser beams relative to the atomic beams are exaggerated in the figure
(adopted from Fig.1 of [88] with permission).

in the state $3^2 S_{1/2}$, $F = 2, m_F = 2$, which allows to make transition to the state $3^2 P_{3/2}$, $F = 3, m_F = 3$ (corresponding to D_2 line of sodium) by interacting with a circularly-polarized laser light. Two different types of cavities were used in the experiments, i.e., the confocal ring cavity and out-of-confocal standing-wave cavity [88]. For the confocal ring cavity, the separations of mirrors was 5.0 cm, having a free spectral range (FSR) of 1.5 GHz. As this separation of mirrors changed to 5.5 cm, it formed a standing-wave cavity with FSR of 2.7 GHz. These cavities sustained a single mode having waist of 69 μm. The mirrors had transmission coefficients $T_1 = T_2 = (3.0 \pm 0.1) \times 10^{-3}$ [5,88]. The Finesse (F) of the confocal ring cavity and standing-wave cavity were 263 and 650, respectively. The dye laser having a linewidth of 500 kHz was employed in the experiment and the laser was both intensity and frequency stabilized. The fundamental TEM$_{00}$ mode of the cavities was mode-matched with the laser beam with about 94 percent efficiency [5,88]. The output transmitted power through the cavity was monitored employing a calibrated photomultiplier tube (PMT). Small-signal absorption measurement was done before each experiment. For experiments demonstrating mixed absorptive-dispersive AOB [88] under the off-resonant conditions of the laser frequency from the atomic line center frequency, the laser frequency was measured without any interruption employing a separate interferometer. In this way the continuous recording of the laser frequency detuning with respect to the atomic line center frequency was made possible. In order to make quantitative comparisons with theory, experimental parameters were measured carefully with absolute values. Such quantitative measurements were not available in initial AOB experiments. The atomic frequency detuning was estimated by measuring the calibrated line profile of optically-pumped fluorescence. The value set for the atomic detuning was within ± 15 MHz from the line center. Controlling of the cavity frequency detuning was obtained by varying the applied voltage to a PZT which changed the position of the cavity input mirror M1. The experimental measurement of cavity detuning was achieved by comparing the slope of the input-output response high on the upper branch of AOB hysteresis curve, with the resonant empty cavity slope [5,87,88]. In this way one could directly measure the cavity detuning, assuming it was done on a sufficiently high lying part of the upper branch and did not depend on calibration factor of detector and also allowed to know the sign of the detuning [87,88].

The atomic density in the experiments was altered through changing the

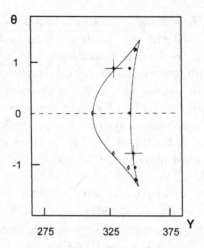

Fig. 2.9 Normalized input switching intensities Y_1 (rhombus) and Y_2 (dots) for the standing-wave cavity as a function of cavity detuning Θ with no atomic detuning $\Delta = 0$ and $C = 15$. The solid line is the theoretical prediction for Y scaled down by a factor of 0.91, which is within experimental uncertainties (reprinted from [88] with permission).

temperature of the oven slowly and the modulation frequency for input intensity was 50 Hz. This particular scan rate so selected was to achieve the steady-state adiabatic behavior. The output and input powers (P_t versus P_i) were monitored continuously by two PMTs and displayed on an X-Y oscilloscope. Atomic and cavity detuning parameters, as well as other ones, were varied during the experiments. A comparative study of normalized input switching intensity [88] as a function of cavity detuning parameter is given for the standing-wave cavity case in Fig.2.9. In this figure an absolute comparison has been displayed (without fitting parameters) between the experimentally measured values and theoretically calculated results for input switching points as a function of cavity detuning (Θ), while keeping atomic detuning fixed in the standing-wave cavity.   The hysteresis loop for pure absorptive AOB was also easily observed in this system, which is shown in Fig.2.10. The curves in this figure are for the incident power vs transmitted power when atomic beams were absent (Fig.2.10(a)), and present (Fig.2.10(b)), respectively [87]. The lower threshold as well as upper threshold of the bistability were estimated for many hysteresis curves of AOB under distinct parametric conditions, which are plotted in Fig.2.11 as a function of C_e (the atomic cooperativity parameter as defined in [87], but it is same as C, defined after Eq.(2.12)). These curves also provide relative uncertainties of both switching powers and data determining C_e.

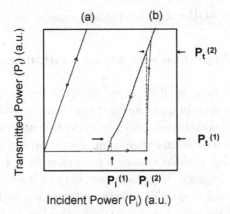

Fig. 2.10   x-y oscilloscope traces of transmitted laser power P_t vs incident laser power P_i for zero atomic and cavity detunings. (a) No intracavity atomic beams. (b) Intracavity beams with resonant absorption $\alpha l = 1.13$ and with $P_i^{(1)} = 70~\mu$W, $P_i^{(2)} = 104~\mu$W (reprinted from [87] with permission).

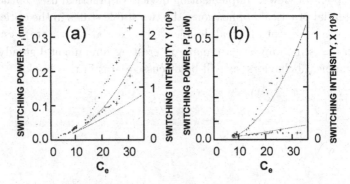

Fig. 2.11   (a) Incident switching power $(P_i^{(1)}, P_i^{(2)})$ vs effective atomic cooperativity C_e as determined from Eq.(1) of Ref.[87]. (b) Transmitted switching power $(P_t^{(1)}, P_t^{(2)})$ vs C_e. The overall determination of C_e is uncertain to ± 15 %. The curves are based on theoretical results of Eq.(3) (Ref.[87]) and plotted relative to the right-hand (intensity) scale (adopted from [87] with permission).

The results obtained from these experiments were also compared with the Gaussian-beam theory of AOB and found to be in very good agreements [87]. Detail comparisons of other experimental parameters in terms of phase diagram in the (C,Θ) plane were also carried out in these works [87,88]. Such quantitative comparisons between experimental measurements and theoretical calculations have provided a good understanding of the absorptive and dispersive AOB in two-level atomic system.

2.7   Potential Applications of Atomic Optical Bistability

The controlling of light by light is one of the most intriguing and interesting fields of research in quantum and nonlinear optics. There can be a large number of potential applications in optical sciences and related fields, such as all-optical switches in optical communication and signal processing. For example, finding suitable materials to implement optical transistors in various frequency regimes, all-optical switching in different frequency domains, optical-atomic memories to store information and quantum information processing using such memory devices and so on. As shown in the above described theories and experiments, the optical bistability behaviors are very sensitive to the absorption, dispersion and nonlinear properties of the intracavity medium. Therefore, studies of AOB in various nonlinear optical systems, especially those displaying EIT, are very important from the point of view of implementing devices of practical uses. Other than the potential practical applications, the two stable states in the optical bistability also represent an ideal double-well potential (two-state) system, which can be used to investigate many interesting fundamental steady-state and dynamical behaviors, as will be discussed later in this book.

Chapter 3

Three-level Atoms as the Intracavity Medium and Atomic Optical Bistability

3.1 Three-level Atoms as the Intracavity Medium

It is possible to enhance the atom-field interaction using an optical cavity because of the multiple passes through the intracavity medium and longer life-time of the photons inside such cavity. Thus it is advantageous to use optical cavities for ultra-sensitive spectroscopic detection purposes, as well as to enhance nonlinear optical processes [62]. The presence of atomic medium inside an optical cavity also modifies its transmission properties due to the absorption and dispersion caused by such intracavity medium, as mentioned in subsection 1.4.3, and sections 2.1 and 2.2. There are several distinct advantages of using multilevel atoms as intracavity medium instead of two-level atoms, and hence during last two decades many interesting experiments in nonlinear optics were reported using multilevel atomic systems. In such experiments having multilevel atoms interacting with coherent fields, significant modifications of absorptive and dispersive properties of a weak probe field near one atomic transition were observed due to the presence of a stronger field interacting with a nearby transition [18-25] which is better known as EIT has been discussed in detail in Chapter 1. Such modified absorption and dispersion properties are due to the generation of atomic coherence and quantum interference because of the simultaneous presence of the probe and coupling fields in the three-level atomic systems, which is not available with the two-level atomic systems. These EIT systems have also displayed several other interesting effects such as the group velocity reduction of light to a few meters per second [25,73,74,77], lasing without inversion [89], enhanced index of refraction [90], waveguides written optically in atomic vapor [91], and storage of photon [92]. When such coherently-prepared atomic (i.e., three-level EIT) medium is placed inside an optical cavity the atomic fluorescence spectrum and cavity transmission profile will also be get altered due to the significant changes in the linear absorption and dispersion properties of the intracavity medium [93]. The substantial changes in optical properties of three-level atomic systems, including enormous reduction in absorption of the probe (cavity) laser field, enhanced sharp linear dispersion, as well as enhanced Kerr nonlinearity dispersion resulted in further explorations of cavity linewidth narrowing and frequency pulling [94] and AOB [37]. Cavity linewidth narrowing was experimentally demonstrated in an optical ring cavity containing three-level Λ-type rubidium atoms [95]. The sharp dispersion feature observed in the three-level EIT medium is due to the presence of induced atomic coherence and quantum interference near EIT resonance [25]. The controls of cavity linewidth with the coupling beam intensity and atomic density were experimentally demonstrated. These observations can be utilized in carrying out precision measurements in relation to the spectroscopic work as well as in nonlinear optical experiments at very weak light levels [96,97]. When EIT medium is placed inside an optical cavity, the conditions for realizing cavity ring-down effect can be easily fulfilled due to the reduced absorption and enhanced dispersion, which has been discussed in subsection 3.1.2 of this chapter. The enhanced Kerr nonlinearity in EIT medium makes the observation of AOB much easier with lower thresholds [42] and has been described in subsection 3.2.2. Controls of the shapes and rotation of the hysteresis cycles in such three-level AOB systems have been achieved [44] by using the coupling laser beam frequency detuning and intensity and is covered in subsection 3.2.3.

3.1.1 Cavity linewidth narrowing effect due to three-level medium inside an optical cavity

3.1.1.1 Theoretical calculations

We begin by theoretically describing frequency pulling and cavity linewidth narrowing effects of a three-mirror optical ring cavity containing three-level Rb atoms under EIT conditions. A three-mirror optical ring cavity containing an atomic medium has been shown in Fig.1.8. The optical path length of this cavity is 'l'. The cavity length can be scanned with the help of a PZT, on which a mirror with reflectivity of near 100% is mounted. The intensity transmissivity T (reflectivity R) of the input mirror M2 is assumed to be identical to that of the output mirror M1 with $T = 1 - R$. We rewrite the expression of cavity transmission given in Eq.(1.37) in a slightly different form in terms of the cavity response function $S(\omega_P) \equiv I_T/I_0$ of an empty optical ring cavity to a probe (cavity) laser field of frequency ω_P as [62]

$$S(\omega_P) \equiv I_T/I_0 = \frac{T^2}{(1-R)^2 + 4R\sin^2(\delta/2)},\tag{3.1}$$

in which δ_p is the phase detuning for an empty cavity goes as

$$\delta_p = \frac{\omega_P l}{c} - 2\pi m,\tag{3.2}$$

where m is an integer. The cavity-laser resonance condition is satisfied when $\delta_p = 0$. Under this condition the cavity output intensity profile shows a maximum. The Finesse (F) of the cavity is defined as the ratio of the FSR of cavity to the FWHM of the cavity output transmission profile. The explicit expression of Finesse is given by $F = \frac{\pi\sqrt{R}}{(1-R)}$ (see Eq.(1.41). In the presence of an atomic medium inside the optical ring cavity, the transmission properties of the cavity given in Eq.(3.1) modify. This is due to the effect of the absorptive and dispersive properties of the medium on the transmission properties of the cavity. If the medium placed inside the cavity is extremely absorptive at cavity resonant frequency then the transmission from cavity is completely suppressed. When the dispersion is present in the cavity, i.e., $dn/d\omega_P \neq 0$, the atomic medium produces frequency shifting effects in the cavity. The modified cavity response function (Eq.(3.1)) of the three-mirror optical ring cavity with an atomic medium inside a vapor cell of length L kept in one side arm of the cavity is given by (see Eq.(1.48))

$$S(\omega_P) \equiv I_T/I_0 = \frac{T^2\kappa}{(1-R\kappa)^2 + 4R\kappa\sin^2(\delta/2)},\tag{3.3}$$

and δ_p modifies to

$$\delta_p = \frac{\omega_P l}{c} + \frac{(n-1)\omega_P L}{c} - 2\pi m.\tag{3.4}$$

The atomic absorption is included through the term $\kappa \equiv \exp[-\alpha L]$, where the parameter α is called the absorption constant. The relationship between resonant frequency ω_a of the cavity containing an atomic medium to the resonant frequency ω_c of the empty cavity is given by

$$\omega_a = \frac{\omega_c}{1 + \frac{L}{l}(n-1)},\tag{3.5}$$

in which $n \equiv n(\omega_a)$ represents refractive index of the atomic medium for the probe laser (cavity field) at the resonant frequency ω_a. It is not possible to calculate the resonant frequencies for the cavity filled with medium because Eq.(3.5) is not an explicit equation for ω_a. Under the condition $n \sim 1$, the resonant frequency ω_a of the cavity containing an atomic medium is equal to the resonant frequency ω_c of the empty cavity. This will be true also for the situation when ω_a is equal to ω_{21}, where ω_{21} is the atomic transition frequency and $n \sim 1$. In other words when $\omega_a = \omega_{21}$ then that means $\omega_c = \omega_{21}$ for this situation. After some calculations it is easy to show that the ratio of the medium-filled cavity linewidth $\delta\omega$ to the empty cavity linewidth C can be represented by the following expression [62]

$$\frac{\delta\omega}{C} = \frac{1 - R\kappa}{\sqrt{\kappa}(1 - R)}\left[1 + \frac{L}{l}[n(\omega_a) - 1] + \omega_a\frac{L}{l}\frac{\partial n}{\partial \omega_P}|_{\omega_p=\omega_a}\right]^{-1}. \qquad (3.6)$$

Over the cavity linewidth the value of κ is considered to be a constant. Under the resonance condition of the cavity and atomic medium, e.g., $\omega_a = \omega_{21}$, the index of refraction at $\omega_P = \omega_a$ is equal to 1 and so the value of the quantity $\Delta\omega/C$ is decided by the dispersion slope $dn/d\omega_P$ at the atomic transition frequency ω_{21}. For the negligible absorption, i.e., $\kappa = 1$ and with a dispersion slope in the order of 10^{-12} s, $l = 40$ cm, $L = 5$ cm, and $\omega_{12} = 10^{15}$ Hz, the ratio $\frac{\delta\omega}{C}$ becomes 0.5 [62]. The cavity linewidth narrowing is not so prominently visualized for the two-level atomic medium inside an optical cavity because (a) the absorption is high, and (b) dispersion is not steep enough near the atomic transition frequency and it has got an anomalous character. Consequently, in practical situation, the cavity linewidth narrowing is hard to be observed for an intracavity two-level atomic medium. However, three-level EIT medium, as shown in Fig.1.9(b), is capable of exhibiting a substantial absorption reduction [4,24], and a steep dispersion change near the atomic transition frequency, which are the essential requirements to observe cavity linewidth narrowing in an appreciable amount.

The three-level atomic system under consideration is shown in Fig.1.9(b). In this system a probe laser with weak intensity having Rabi frequency $\Omega_P = \mu_{21}E_P/\hbar$ interacts with the atomic transition $|1 >$ and $|2 >$, while another laser called coupling laser having stronger intensity with Rabi frequency $\Omega_C = \mu_{23}E_P/\hbar$ interacts with the transition $|3 >$ and $|2 >$. The quantities E_P and E_C are characterizing the field strengths of the probe and coupling lasers, respectively, and μ_{21} and μ_{23} are the corresponding dipole moments for the atomic transitions. In order to eliminate first-order Doppler effect, the probe and coupling lasers propagate collinearly through the three-level Λ-type system as discussed in subsection 1.3.2 of

Chapter 1. The complex susceptibility for this system, averaged over a Maxwell-Boltzmann velocity distribution (Eq.(1.31)), is given by Eqs.(1.59) and (1.60). Note that in the expression of χ in Eq.(1.59), $N_0 = N_0(T)$ is the temperature-dependent atomic density function and atomic velocity $u = \sqrt{2k_B T/M}$, $\gamma = \gamma_{21} + \gamma_{23} + \gamma_{31}$, where $\gamma_{21} + \gamma_{23}$ provides the total radiative decay rate from level $|2>$ and γ_{31} is the decay constant for the all nonradiative decay from level $|3>$. Since the lasers posses some finite linewidths which can be added in the decay rates and the effective decay rates are given as $\gamma \to \gamma + \delta\omega_P$ and $\gamma_{31} \to \gamma_{31} + \delta\omega_P + \delta\omega_C$, where $\delta\omega_P$ and $\delta\omega_C$ are the half-linewidths of the probe and coupling lasers, respectively [4,24,62]. The frequency detunings of the probe and coupling lasers are given by $\Delta_P = \omega_P - \omega_{21}$ and $\Delta_C = \omega_C - \omega_{23}$, respectively. It is easy to have the following relation

$$n \approx 1 + \chi'/2, \tag{3.7}$$

in which χ' comes from $\chi = \chi' + i\chi''$. The expression for linewidth narrowing, $\Delta\omega/C$ can be derived using Eq.(3.6). When EIT conditions are fulfilled in the three-level Λ-type ^{87}Rb atomic system, the normal linear dispersion produces a steep slope for the probe detuning Δ_P of up to around 7 MHz [25]. The expression of Eq.(3.6) for the cavity linewidth narrowing can be simplified in this region [62,94] as

$$\frac{\delta\omega}{C} = \frac{1 - R\kappa}{\sqrt{\kappa}(1 - R)}[1 + \eta]^{-1}, \tag{3.8}$$

where $\eta \equiv \omega_a(L/2l)(\partial\chi'/\partial\omega_P)|_{\omega_P=\omega_a}$. The roles played by the absorption and dispersion of the atomic medium are opposite in the cavity linewidth narrowing phenomenon and thus quite intriguing. For any appreciable narrowing, high dispersion and low absorption are required. The effects of dispersion and absorption in the cavity induced line narrowing can be understood by following considerations. In Fig.3.1 the cavity response function $S(\omega_P)$ is plotted with respect to the probe detuning Δ_P, ignoring absorption efects and considering that κ is a constant, i.e., $\kappa = 0.517$ with the coupling beam power of $P_C = 1$ mW at a vapor cell temperature of $T = 87\ ^0$C in the experiment for the entire range of ω_P values. Ideally the requirement ω_a to be equal to the atomic transition frequency ω_{21} need to be fulfilled and that can be achieved by small change in the cavity length l. Also, the coupling field detuning Δ_C is set to be zero. It is quite easy to observe that the empty cavity resonant peaks are moved towards the atomic transition frequency ω_{21}. Here, absorption is considered to be a constant for all probe frequencies which obviously is an oversimplification since in real experiments that is not the case due to various atomic resonances [62]. In Fig.3.2(a) the effect of frequency-dependent absorption profile with EIT (under different frequency scale) is shown under the similar parametric conditions. Hence it is easy to understand that only the center peak of the cavity transmission will exist due to the narrow EIT transmission peak right in the center of the ω_{21} absorption line profile and other nearby peaks will be absorbed by the large absorption outside the EIT window [62]. In Fig.3.2(b) the refractive index of the three-level atomic medium is plotted as function of Δ_P (dispersion curve) under the similar parametric

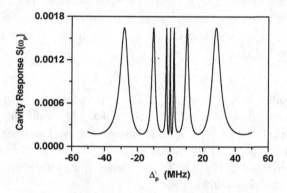

Fig. 3.1   Theoretically calculated cavity response as a function of probe detuning. $\kappa = 0.517$ is assumed to be a constant for all frequency values. $T = 87\ ^0$C, $P_c = 1.0$ mW, $\Delta_c = 0$, $R = 0.98$, $l = 36.5$ cm (reprinted from [62] with permission).

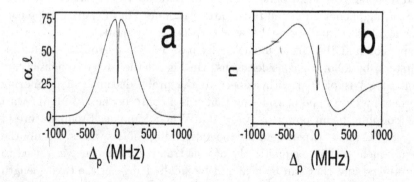

Fig. 3.2   Theoretically calculated absorption and dispersion curve for of a 5-cm-long Rb vapour cell showing (a) characteristic EIT transmission spike and (b) sharp dispersion slope at $\omega_p = \omega_{21}$, $T = 87\ ^0$C, $P_c = 1.0$ mW, $\Delta_c = 0$ (reprinted from [62] with permission).

conditions as used in Fig.3.2(a). The dispersion curve shows a sharp slope near the probe resonant frequency ω_{21} as described in subsection 1.5.3 (A) and in Ref.[25]. Note that in the theoretical model of the susceptibility equation, the probe intensity has always assumed to be extremely weak [24]. With a very low probe intensity, despite the reduced absorption due to EIT at the transition frequency ω_{21}, the residual absorption turns out to be quite high to measure any transmitted field at the cavity output. When the temperature of the atomic vapor cell is increased, the atomic density in the atomic vapor rises rapidly, which results in a stronger absorption of the probe field. In order to get an appreciable cavity output, the increase of probe intensity is normally required, which brings a saturation effect in the absorption of the probe beam. This means that consideration of a probe intensity beyond the weak-probe limit is required, so the equation for obtaining the susceptibility can not provide accurate values of κ that are in good agreement with the experimental results measured at higher intensities of probe beam [62,95].

In the experiment which will be described in the subsequent subsection below, the cavity linewidth narrowing ratio $\Delta\omega/C$ was measured at different values of coupling laser power. The cavity linewidth narrowing crucially depends on the atomic dispersion slope $dn/d\omega_P$, so the behavior of this dispersion slope at $\omega_P = \omega_a = \omega_{21}$ and its dependence on the coupling beam power need to be investigated. It is easy to study $dn/d\omega_p$ as a function of the coupling beam power using Eq.(1.59) and Eq.(3.7), as shown in Fig.3.3, at a vapor cell temperature of 87 ^{0}C. With the increase of coupling power from zero the dispersion slope increases rapidly at the beginning, but it reaches a maximum value when the coupling beam power is at the level of 1 mW. Further increasing the coupling power causes the dispersion slope to go down gradually. This decrease in dispersion slope at higher coupling power is a consequence of power broadening in the EIT transmission profile. The reduced dispersion slope gives rise to a broader cavity transmission profile. Thus for a particular atomic density, a maximum cavity linewidth narrowing can be achieved under the optimum condition of the coupling beam power [62].

3.1.1.2 *Experimental Investigations*

The experimental arrangement used for the demonstration of cavity linewidth narrowing was similar to Fig.1.15 [95]. The lasers used for the

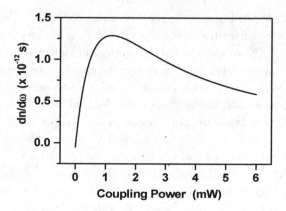

Fig. 3.3   Theoretical plot of dispersion slope at ω_{21} as a function of coupling power. $T = 87\ ^0\mathrm{C}$, $\Delta_c = 0$ (reprinted from [62] with permission).

probe and coupling transitions were current and temperature stabilized single-mode tunable Hitachi HL7851G diode lasers. The stabilization of probe laser frequency was also provided through the weak feedback from a grating mounted in the Littrow configuration [62,98]. The half-linewidths of these lasers were $\delta\omega_P \simeq 2\pi\times 1$ MHz (probe) and $\delta\omega_C \simeq 2\pi\times 2.1$ MHz (coupling), respectively. About 10% of the probe and coupling laser beams were introduced into an auxiliary Rb cell with the help of polarizing beam spitters PBS1 and PBS2 and then these beams were allowed to pass through a FP cavity detected by a detector so that the laser frequencies could be monitored. Another beamsplitter PBS4 was used for combining remaining parts of the laser beams which were then focused into a Rb vapor cell wrapped by μ-metal sheet for magnetic shielding. As mentioned earlier, the Rb vapor cell was 5-cm long with Brewster-cut windows and a heating tape was wrapped over the μ-metal sheet for controlling the temperature of the cell to change the atomic number density. The three-mirror optical cavity had a flat output mirror M1 (transmissivity of approximately 1%) and a concave input mirror M2 (transmissivity of approximately 3%). The third cavity mirror was a high reflector and mounted on a PZT driver, as shown in Fig.1.15 [95]. The radius of curvature of these concave mirrors were about 10 cm. The length and Finesse of the empty cavity were 36.5 cm and 100, respectively. The measured FSR of the empty cavity was 822 MHz leading to the empty cavity linewidth C to be $2\pi \times 8.22$ MHz. The probe and coupling beams were allowed to propagate in the same direction through the atomic vapor cell and were orthogonally polarized. Two lenses with focal lengths of 15 cm and 40 cm were used for focusing the probe and coupling beams, respectively, into the optical cavity. The injected probe beam circulated inside the cavity and the coupling beam was introduced from PBS4 with the single pass through the atomic vapor cell, as shown in Fig.1.15. The diameters of probe and coupling beams at the center of the Rb cell were estimated to be around 140 μm and 280 μm, respectively. The Finesse of the cavity was reduced down 51 approximately after introductions of the Rb cell and PBS4 in the cavity, for the probe frequency far off resonance. The reduction in Finesse was due to the surface reflection losses from the atomic vapor cell and PBS4 [62]. The experiment was started by removal of the mirror M2 and the probe laser frequency ω_P was scanned near the atomic transition frequency ω_{21}. The frequency range (in terms of angular frequency) of scan was about $2\pi \times$ 770 MHz and scan time was of 8.5 ms using a zig-zag current scanning waveform, which had a peak-to-peak voltage of 5 mV. At this stage the coupling laser was turned on and its frequency was tuned near to the atomic transition frequency ω_{23}. If the two frequencies matched exactly, i.e., $\omega_C = \omega_{23}$, the characteristic EIT transmission spike in the ω_{21} absorption profile appeared right in the center. After the preliminary adjustment of the laser frequencies, mirror M2 was placed back to realize the making of optical ring cavity. The next step was to tune the optical ring cavity with the probe atomic transition, i.e., to achieve the condition $\omega_{21} = \omega_a$. This condition was realized by

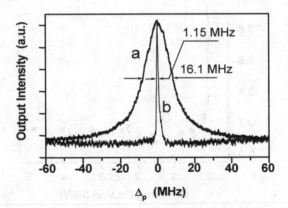

Fig. 3.4   Intensity of cavity output versus probe frequency detuning for: (a) No coupling beam and the probe frequency well outside the absorption line, and (b) Coupling beam power of 0.81 mW and probe frequency scanned through the probe transition ω_{21} keeping $\Delta_c = 0$. Note: the intensity scales for (a) and (b) are different. $T = 87\ ^0$C (reprinted from [62] with permission).

adjusting the length l of the cavity by changing the driving voltage to the PZT, where mirror M3 was mounted on. Under this condition, the cavity transmission peak got maximum in height and became extremely narrow. In Fig.3.4 an experimental comparison is given for two cavity transmission peaks, one for the probe beam at a frequency far from ω_{21} without coupling beam (Fig.3.4, curve a) and another for the probe beam on-resonance with a coupling beam power of 0.81 mW (Fig.3.4, curve b). The Rabi frequencies associated with the probe and coupling laser fields at the center of the Rb vapor cell were $\Omega_P = 2\pi \times 4.9$ MHz and $\Omega_C = 2\pi \times 49.8$ MHz, respectively. The measured linewidth of cavity transmission profile in the presence of Rb vapor cell and the PBS4 but in the absence of the coupling laser was found to be $C' - 2\pi \times 16.1$ MHz. The measured value of this linewidth changed to $\Delta\omega = 2\pi \times 1.15$ MHz when the coupling laser was present forming EIT configuration. The linewith reduction was found to be about a factor of 14 in comparison to this C'. In comparison to the empty-cavity linewidth C, the factor for cavity linewidth narrowing was of the order of 7 [62,95]. The dependence of the dispersion slope $dn/d\omega_P$ on coupling power was discussed in the last subsection (Fig.3.3). According to the theoretical model, by increasing the coupling laser power, there is a drastic increase and then slow drop-off in the dispersion slope that could greatly alter the amount of cavity linewidth narrowing. In Fig.3.5, experimentally measured ratio $\Delta\omega/C$ with the variation of the coupling power P_C is depicted. In the the-

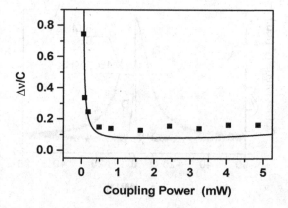

Fig. 3.5   Ratio of narrowed linewidth $\Delta\omega$ normalized to the empty cavity linewidth C measured as a function of coupling power for $T = 87$ ^{0}C and $\Omega_P = 2\pi \times 4.9$ MHz at the centre of the atomic vapor cell. The theoretical curve assumes $\gamma = 2\pi \times 4.0$ MHz, $\gamma_{31} = 2\pi \times 3.2$ MHz, and the diameter of the coupling beam at the beam waist is 280 μm (reprinted from [62] with permission).

oretical model, probe saturation effect was not considered so the simulated curve was plotted using an empirical fit to the experimentally measured value of κ [95]. The values obtained experimentally match reasonably well with theoretical predictions and show a maximum linewidth narrowing at about 1-2 mW for the coupling power. The narrowing gradually decreases at high power level of the coupling laser due to power broadening.

The cavity linewidth narrowing ratio $\Delta\omega/C$ was also measured as a function of atomic density or temperature of the vapor cell, which is shown in Fig.3.6 along with a theoretically simulated plot. In these measurements the power of the coupling laser was about 1 mW. The theoretically simulated curve (depicted in inset of Fig.3.6) was obtained using a numerical fitting to the experimentally measured values of κ as a function of temperature. The cavity linewidth gets narrower with the increase of temperature since EIT in the Rb vapor becomes more prominent as atomic densities go up. In order to see better cavity linewidth narrowing, high temperature of the Rb vapor cell is required so that steep slope could grow in the dispersion curve near ω_{21}. In the experiment the minimum temperature required for observing cavity linewidth narrowing was about 40 ^{0}C. The modification of the cavity transmission profile is greatly affected by the atomic absorption κ and the dispersion slope $dn/d\omega_P$, as indicated by Eq.(3.8). Usually, ab-

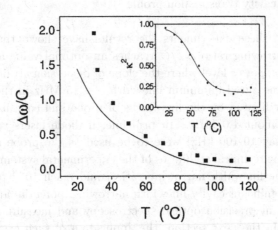

Fig. 3.6 Cavity linewidth narrowing ratio as a function of temperature (atomic density). Experimental values were measured for a coupling power of 1.0 mW. The theoretical curve was calculated assuming $\gamma = 2\pi \times 4.0$ MHz, $\gamma_{31} = 2\pi \times 3.2$ MHz, and the diameter of the coupling beam at the beam waist is 280 μm. Insert: κ^2, measured versus temperature for a 1.0 mW coupling beam (reprinted from [62] with permission).

sorption causes suppression of cavity transmission and results in a broader cavity transmission profile. However, the highly dispersive atomic EIT medium with a steep dispersion slope can induce frequency pulling and cavity linewidth narrowing effects [94,95]. The conditions provided by the three-level EIT medium, i.e., reduced absorption and enhanced linear dispersion, are not met in a two-level atomic system because in the two-level system absorption near atomic transition frequency is very high and the dispersion slope is small and anomalous, which broadens the cavity transmission profile.

In the above discussion, the linewidth narrowing of cavity transmission profile has been attributed to the modified dispersion and absorption properties of the intracavity EIT medium (Eq.(3.8)) [94,95]. Actually, such cavity linewidth narrowing phenomenon can be easily understood by the following consideration. The cavity linewidth narrowing is inversely proportional to the lifetime of photon inside the cavity [62]. The lifetime of photons can be increased by making them travel slower in the cavity, so they will not hit the 'lossy' mirrors as often (to escape from the cavity). In three-level EIT medium, the achieved sharp normal dispersion slope greatly reduces the speed of photons inside the cavity [25]. Consequently the lifetime of photons in the cavity increases, which leads to narrowing of the linewidth in cavity transmission profile.

Note that in these experiments the results have shown that the cavity linewidth narrowing ratio $\Delta\omega/C$ reaches an optimal value almost at the same coupling power level where the slope of dispersion attains the largest value. The measured minimum linewidth of 1.15 MHz in the experiment was basically limited by the linewidths of probe and coupling laser used, which were about 1.0 MHz. In principle, if diode lasers with narrower linewidths (say 10-100 kHz) were to be used as the probe and coupling lasers, and the technical stabilities of the experimental system were further improved, the linewidth narrowing ratio could reach 10^{-3} to 10^{-4} level. With such significant cavity linewidth narrowing, potential applications of such scheme in precision optical spectroscopy and measurements can be envisioned. In the next section, the application of such cavity linewidth narrowing in cavity ring-down spectroscopy (CRDS) will be discussed.

The above cavity linewidth narrowing has been realized with an enhanced normal dispersion slope in the EIT system. Another interesting study could be with the steep anomalous dispersion slopes [99,100], i.e. $dn/d\omega_P < 0$ for the atomic medium inside the optical cavity. Under such condition the cavity linewidth $\Delta\omega$ will become broader to start with and then it will become narrower again as $dn/d\omega_P$ goes more negative, making $\Delta\omega/C < 0$, which are clearly revealed by Eq.(3.6) or Eq.(3.8). Physically, a negative cavity linewidth narrowing ratio $\Delta\omega/C < 0$ should mean a switching in frequency components above and below the reference frequency in the definition of the cavity linewidth [62]. The broadening of cavity linewidth (so called the 'white-light cavity') by anomalous dispersion of the intracavity medium was experimentally demonstrated [101] using double gain peaks in a four-level atomic system [99]. Another way to realize total anomalous dispersion has been to use the nonlinear dispersion slope, which is anomalous [33], to compensate the normal linear dispersion slope. At higher input probe intensities, the total dispersion inside cavity due to the EIT medium can be made to be anomalous, therefore, demonstrating cavity linewidth broadening [102] and 'white-light cavity' [103]. Such broadened cavity linewidth is a result of fast moving photons inside the cavity with a short cavity lifetime.

By investigating the role of an intracavity atomic medium on the cavity transmission profile, better understanding can be gained into the effects that the absorption and dispersion, as well as the nonlinearity, of the intracavity medium create near cavity resonance. Resonant frequency pulling of cavity mode and cavity linewidth narrowing/broadening are critically determined by the atomic absorption as well as dispersion slope [62,94,95,102]. The role of optical cavities in precision spectroscopy and laser research is quite significant. So, the understanding of the influences of the intracavity dispersion and absorption properties on the cavity transmission profiles is very important. The ability to control these properties with the help of induced atomic coherence and quantum interference effects can give new methods for studying novel optical phenomena and identifying other interesting new applications. One of such examples is given in the following subsection.

3.1.2 *Enhanced cavity ring-down spectroscopy with a three-level electromagnetically induced transparency system*

The original motivations for the development of cavity ring-down spectroscopy (CRDS) are ring gyroscopes and interferometers where high reflectivity mirrors are used. The basic concept behind the cavity ringdown technique can be found in several pioneering works made in the early eighties, which were aimed at characterizing the reflectivity of high reflecting mirror surfaces for several different applications including aerospace industry [104,105]. The innovation of the optical ring gyroscope functioning at the He-Ne laser wavelength, 632.8 nm, gave a tremendous impact to aviation and navigational systems [104]. In the navigational gyro, the accuracy was attributed to the optical residence time within the ring cavity. This residence time could be increased by enhancing the reflectivity of the mirror surfaces with appropriate coating material. High reflectivity mirrors with reflectivities exceeding 99.9% gave great impact to the optical experiments. Mirrors with such coatings were also of significant importance in interferometric applications. During that early period, to characterize mirror coating was a challenging task and there was uncertainty in estimating any change below 0.1 percent. So the improvement in incremental coating quality of mirrors was extremely difficult. Soon a new method based on phase shift measurement technique came into existence which provided an accurate methodology to measure average reflectivity within 0.01 percent for a pair of mirrors forming a cavity [104]. Another method involving a shuttered laser injection of an optical cavity was discovered. Using this method the reflectivity of a mirror pair to an accuracy of 0.0005 percent was determined [104,105]. All these methodologies led to the development of CRDS. In the spectroscopic studies the CRDS was first time utilized to study the molecular absorption measurements [106,107] in late eighties. Since then many innovations came out and this technique has been established as the ultra-sensitive spectroscopic detection technique in the optical absorption analysis of atoms, molecules, and optical components [108]. In particular, it has been used in measuring ultrasensitive direct absorption [107,109], exploring the chemical kinetics and absorption bands of molecules [110,111] and in concentration measurements of minor species in flames [112]. The CRDS technique provides a great amount of sensitivity in detection of species within optical cavity. The CRDS is quite simple to use and does not require any elaborated equipment or instrument to carry out measurements. As we will see in the following discussion that the theory behind operation of CRDS is also very simple.

The fundamental idea of CRDS technique is to do a measurement of the light decay in optical resonator possessing a high Finesse. The optical cavity is formed by a pair of extremely good quality mirrors separated by a distance l, usually one of the mirrors is a plane mirror and the other one is a concave mirror. The reflectivities (R values) of these mirrors are typically required to be greater than 0.9999. The simplest way to describe this technique is as follows: a pulse of laser light goes into the optical cavity through one of the mirrors, and the little amount of light coupled through the mirror into the cavity is reflected back and forth several times between the two mirrors of the cavity. A small amount of light gets lost during each travel in the cavity due to transmission/absorption by the mirrors or absorption by the medium inside the cavity. The overall effect of all these processes is that there is an exponential decay of light intensity in time inside the cavity. This decay of light can be recorded by suitably positioning a sensitive photodetector behind the second mirror, from which a small amount of light is transmitting out of the cavity during each travel. The decay constant τ of light is given by [106]:

$$\tau = \frac{l}{c(\beta + \alpha L)}, \tag{3.9}$$

where l is the length of cavity, c is the speed of light, β represents the empty cavity loss which is equal to 1 - R for a two mirror cavity in terms of the reflectivity R of mirror. The absorption coefficient of the sample of path length L contained in the cavity is given by α [106].

From Eq.(3.9) it is clear that the exponential decay constant of light intensity depends on the absorption coefficient of the sample in the cavity, the length of the cavity and the reflectivity of the mirrors. So the cavity containing an absorbing sample provides greater losses for light on each pass, and therefore a smaller exponential decay time constant for the light [106]. Scanning the laser wavelength will lead the exponential decay time constant recorded at each wavelength to be converted into an absorption coefficient, so the absorption spectrum of the intracavity medium could be obtained. The decay constant measurement in such experimental arrangement does not depend on the absolute intensity of light entering the cavity, hence the CRDS is advantageous over the single-pass technique for measuring small absorption because it is not sensitive to uncertainties caused by intensity fluctuations in the pulsed laser used. Sensitivities of detection for particular samples depend on their photo-absorption cross sections at the requisite wavelength, but are usually in the range of parts-per-million to parts-per-billion [106]. The sensitivity of the CRDS technique enhances with the increase in the reflectivity R of the mirrors because the loss per cycle is critically dependent on the parameter $\beta = 1 - R$, and hence the requisite path length through the sample can be adjusted to achieve required sensitivity of detection. To increase the sensitivity in this way, in pulsed CRDS, there is reduction in the overall signal levels as the output light level becomes very low in each pass. To increase the output light levels from the cavity the CW technique is used, in which light is temporarily stored in the cavity such that it reaches the detectable magnitude over an appreciable time period covering many ringdown time parameter. After that the light source is switched off, and then the ringdown decay time constant is measured [106]. Here, the ringdown time constant can be estimated by knowing the time required by the light in the cavity to reach a steady-state intensity value. Another way to measure ringdown time is via the phase shift measurement, experienced by an intensity-modulated CW beam circulating in the cavity. From such phase shift measurements, the cavity ringdown time, and consequently, the absorption spectrum could be obtained [106]. The relationship between the phase shift ϕ and the ringdown time τ is given by $\tan \phi = -\omega_m \tau$. In this expression ω_m represents the frequency by which the light source is being modulated. Another variant of the CRDS technique using CW source is the cavity enhanced absorption spectroscopy (CEAS) [113,114]. The basic philosophy behind this technique is that under continuous tossing of the light inside the optical cavity between its mirrors at a wavelength resonant with one of the cavity modes, the total light intensity accumulating inside such optical cavity is approximately proportional to the ringdown time [106]. Hence it is possible to know the absorption spectrum from the measurement of total time-integrated cavity output intensity as a function of wavelength. The technique of CEAS requires the measurement of absolute intensities and not the decay rates so it is very sensitive to fluctuations in laser intensity during the measurement time.

Next, we look for the operation of the pulsed cavity ringdown technique in more detail. The basic idea of its operation is quite simple (as discussed above) and can be considered as a special case of CW excitation of a cavity. The CW source is now replaced by a tunable pulsed laser source [105]. The continuous wave excitation of a cavity is intricate because the energy within the cavity builds up on time scale same as the cavity decay time provided the two cavity modes remain locked during that time. Under this circumstance, shutting off injected light into the cavity very quickly in comparison to cavity buildup or decay time, the eventual decay of the light becomes exponential, which could be demonstrated quite easily when pulses are used [104]. The schematic diagram of the pulsed cavity ringdown technique is shown in Fig.3.7.

Here an optical cavity comprised of two mirrors (each with a reflectivity R for simplicity) couples to a tunable pulsed laser. The laser mode is spatially mode matched to the optical cavity modes. The cavity mirrors are multilayered dielectrically coated reflectors, which permit a very small amount of light to transmit through. When a laser pulse is shorter than the cavity round-trip time it is easy to understand the traveling back and forth of the injected pulse between the two mirrors of the cavity. During each trip between mirrors, a small pulse amplitude gets transmitted out from the cavity. The amplitude of the transmitted pulse is 'T' times the amplitude of the pulse circulating inside the cavity. If there are no absorbing sample inside the cavity and no scattering losses from the mirrors, then $T = 1 - R$. The rate of optical intensity loss from the cavity is then given by [104]

$$dI/dt = -I \times T \times c/2l, \qquad (3.10)$$

in which c denotes the speed of light and l represents the separation between mirrors or the length of the cavity in this case. Eq.(3.10) can be solved to give

$$I = I(0) \times \exp[-(\frac{tTc}{2l})]. \qquad (3.11)$$

The total round-trip loss [104] in the cavity can be estimated by Γ, where

$$\Gamma = 1 - \exp[-(2l/c\tau)], \qquad (3.12)$$

and τ is the exponential decay time of this signal (see Eq.(3.9)). If there are no other loss mechanisms in the cavity then the measured loss will give rise to the transmission curve of the mirrors making the optical cavity

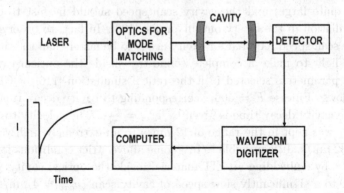

Fig. 3.7   Schematic illustration of the cavity ringdown process (adopted from [104] with permission).

[104]. If there is a resonant absorbing material inside the cavity then the cavity decay rate becomes faster because effective value of Γ increases. It is straightforward to estimate the absorption caused by absorbing sample if the losses due to mirrors are accurately known (which can be determined by measuring the empty cavity). This method provides a straight absorption measurement for very small amount of absorbing species. So a quantitative ultra sensitive absorption measurement can be performed provided one adheres to the standard philosophy of measurements. For example, the bandwidth of the source light injected inside the cavity should be narrower than the linewidth of absorbing sample inside the cavity [104].

Alternatively, CRDS can be achieved in the following way. When an optical cavity is coupled to a monochromatic optical field and the cavity length is scanned faster than the cavity round-trip time of the light, the cavity output field exhibits amplitude oscillation in its usual light decay curve, known as cavity ringdown effect (CRE) [108,115,116]. Such oscillation is caused by the interference between the original input optical field and the field circulating inside the cavity. By changing the cavity scan speed, oscillation frequency can be controlled. By measuring the damping amplitude and period of oscillation in the transmission curve, one can accurately determine the small absorption and reflectivity of the mirrors [107,109].

There are many practical applications of CRDS or such CRE, as discussed above, but at the same time some serious limitations and strict requirements also exist in this technique. For example, the optical cavity needed to show CRE should be of very high Finesse (i.e., R values of mirrors must be quite large) and the cavity scan speed should be fast to observe the oscillation in the cavity output profile [108]. In fact, to observe the CRE, the scan speed required for cavity needs to be faster than the time required for light to make a complete round-trip inside the optical cavity. Typical parameters selected in a theoretical simulation [116] of CRE were as follows: Finesse $F = 5000$, corresponding to cavity decay time of 1.7 μs. The cavity decay time is given by $\tau_{cav} = \frac{Fl_0}{\pi c}$. The velocity to scan cavity v_{cav} was kept in the range of 12 μm/s to an extremely high value around 5932 μm/s. It is possible to overcome above strict conditions for observing CRE by embedding an EIT material inside the optical cavity, which gives rise to a significantly slow speed of cavity scan ($v_{cav} = 4$ μm/s [108]). In this way CRE is enhanced and its practical applications can become wider. As mentioned previously that an EIT medium can give a sharp dispersion change near its EIT resonance, leading to slowing down of the speed of light inside the optical cavity [25,62,94] and consequently increasing interaction time. An EIT intracavity medium has an advantage to have a large dispersion change without any absorption at exact EIT resonance condition.

In the literature the description of CRE has generally been done by finding the electric field inside the optical cavity at any instant of time by adding up all the components of light wave that passed through multiple reflections in the cavity [108,115,116]. A simplified method of modeling the intracavity electric field is given in the following. Let α_P represents the intracavity field, such that the average photon flow, represented in number of photons per second, is given by $|\alpha_P|^2$. An optical ring cavity (Fig.3.8(a)) considered here is made up of four mirrors. Mirrors M1 and M3 are acting as input and output mirrors both having same reflectivity R for the intensity of light. The other mirrors M2 and M4 are one hundred percent reflectors. The mirror M2 is mounted on a PZT so that the cavity length can be scanned [108]. All the description in the following are also applicable to a three-mirror optical ring cavity, as shown in Fig.1.8. The modification of intracavity (probe) field α_P during a complete round-trip time T_0 is related to three parameters: the driving field α_P^{in}, the cavity decay Γ_{cav}, and the round-trip phase shift Θ_{cav}. The equation of dynamical evolution for the intracavity field can be expressed quantitatively as [108]

$$T_0 \frac{d\alpha_P}{dt} = \sqrt{R}\alpha_P^{in} - \Gamma_{cav}\alpha_P + i\Theta_{cav}\alpha_P. \qquad (3.13)$$

In the situation when an empty cavity is in use the total round-trip phase shift is proportional to the geometrical length of the cavity and can be given through the simple expression $\Theta_{cav} = \frac{2\pi(l_0 + v_{cav}t)}{\lambda_P}$, where the wavelength of the input optical field is λ_P, the initial cavity length is l_0, and the parameter for the scanning speed of cavity is v_{cav}. The expression of Θ_{cav} can be used in Eq. (3.13) to get the closed form analytic solution of $\alpha_p(t)$ [108]:

$$\alpha_p(t) = \frac{\alpha_P^{in}\sqrt{\pi R}}{\sqrt{2iT_0a}} \exp[\frac{(\Gamma_{cav} - iat)^2}{2iT_0a}] \times [\mathrm{erf}\ (\frac{\Gamma_{cav}}{\sqrt{2iT_0a}}) - \mathrm{erf}(\frac{\Gamma_{cav} - iat}{\sqrt{2iT_0a}})], \qquad (3.14)$$

in which the term containing l_0 is ignored assuming that initially the field and cavity are in resonance with each other. The parameter $a = 2\pi v_{cav}/\lambda_P$, and the term erf represents the standard error function. For some specific conditions of parameters, which will be given subsequently in the following discussion, the cavity transmission (output) profile exhibits a clear ringdown oscillation. A typical ringdown oscillation is shown in Fig.3.8(b).

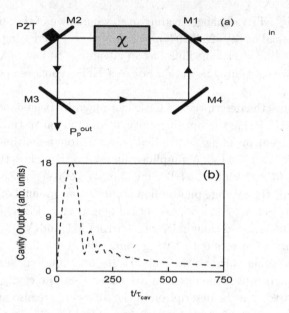

Fig. 3.8 (a) Four-mirror optical ring cavity with an intracavity Three-level atomic medium of type shown in Fig.1.9 (adopted from [108] with permission). (b) A typical cavity ringdown oscillation.

The difference of timing between the first and second oscillation minima is found to be given by $T_{12} \approx 0.5258(l_0\lambda_P/cv_{cav})^{1/2}$ [108]. Similar result can also be obtained from the interference method [115,116].

It is important to investigate the conditions for observing CRE by a substantial scrutiny of Eq.(3.14) and finding its solution numerically. The CRE can be exhibited for $T_0 \sim 1$ ns, provided the condition $v_{cav} \geq 50\Gamma_{cav}^2$ is satisfied; in other words, for a particular cavity decay (related to the reflectivity R), there is a minimum requisite scan speed of the cavity. However, in practice there is a limit on the cavity scan speed due to its mechanical construction and it cannot be increased beyond a specific maximum value. The typical cavity scan speed is of the order of several micrometers per second requiring a cavity decay smaller than 10^{-4}, demanding a cavity Finesse to be of the order of 10^4 [107–116]. There are extraordinary changes in this scenario when an EIT medium is introduced inside the optical ring cavity [108]. A common EIT medium of rubidium atomic vapor in three-level Λ-type configuration has been shown in Fig.1.9(b). A coupling laser (strong) of frequency ω_C near the atomic transition frequency ω_{23} couples levels $|2>$ and $|3>$, whereas a probe laser (weak) with frequency ω_P near the atomic transition ω_{21} couples levels $|2>$ and $|1>$. Only the weak probe laser circulates inside cavity as the cavity field and the coupling laser does not circulate inside the optical cavity. Also, the coupling laser has a different frequency from the probe laser, so there is no direct interaction of this laser with the probe transition of the atomic or molecular system inside the cavity [108]. The real (χ') and the imaginary (χ'') parts of the electrical susceptibility (χ) of EIT medium can be calculated analytically [24] for the probe beam (cavity field). The absorption and dispersion coefficients [108] can be expressed as $\alpha = \omega_P n_0 \chi''/c$ and $\beta = \omega_P n_0 \chi'/2c$, respectively, are plotted in Fig.3.9 with respect to the cavity (probe) field frequency detuning parameter Δ_P. Here, n_0 is the refractive index of the background. The parametric conditions used in obtaining these results are $n_0 = 1$, $\lambda_P = 795$ nm, $\gamma_{31} = 0$, and the coupling laser Rabi frequency $\Omega_C = 20$ MHz. With the EIT medium inside the cavity and the transparency condition is met at resonance, then there is a large change in the dispersion coefficient as discussed in the last subsection. If the decay constant $\gamma_{31} \neq 0$ then some residual absorption at the resonant frequency is expected. Once the EIT conditions is fulfilled for the system, the group velocity of probe beam can greatly reduce down, which can be utilized to increase the photon round-trip time considerably inside the optical cavity, as shown in the last subsection. This eventually leads to narrowing of the cavity linewidth [62,95] and effectively decreasing the decay rate of light inside cavity near the EIT resonance [108]. In the presence of EIT

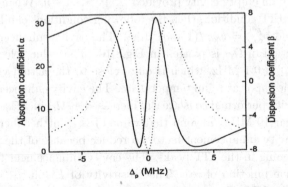

Fig. 3.9   Theoretical calculations of the absorption (dotted curve) and dispersion (solid curve) coefficients of the three-level EIT medium (reprinted from [108] with permission).

medium inside the optical cavity, the photon round-trip time changes to $T^E = T_0(1 + \eta)$, where $\eta = \omega_{21}(L/l_0)(\partial\chi'/\partial\omega_P)|_{\omega_P=\omega_{21}}$ quantifies the enhanced dispersion slope of the EIT medium having length L. At the exact EIT resonant frequency, the nonlinear index of refraction n_2 is zero [33], and consequently the nonlinear phase-shift also goes to zero. It should be noted that the linear phase shift (i.e., the phase shift due to the linear dispersion is also linear) is simply a constant, which shifts the overall cavity transmission profile horizontally. Thus setting the linear phase shift to be zero would not lose the generality of the expression. The equation (Eq.(3.13)) that determines the dynamical evolution of intracavity field, is still effective after the EIT medium is kept inside the optical cavity, provided one uses T^E for T_0. Thus Eq.(3.13) can be recasted as [108]

$$T^E \frac{d\alpha_P}{dt} = \frac{\sqrt{R}\alpha_P^{in}}{1+\eta} - \Gamma_{cav}^E \alpha_P + i\frac{2\pi v_{cav}t}{\lambda_P(1+\eta)}\alpha_P. \qquad (3.15)$$

The overall scaling factor of the output field is changed by the first term of Eq.(3.15) and the new cavity decay Γ_{cav}^E is given by $\Gamma_{cav}^E = \Gamma_{cav}/(1+\eta)$. Clearly in the presence of EIT, the cavity decay factor Γ_{cav} is effectively goes down by $(1+\eta)$. Under the imperfect condition for EIT, the effective cavity decay is governed by $\Gamma_{cav}^E = \Gamma_{cav}/F^e$, where F^e is the cavity enhancement factor, given by $F^e = (1+\eta)\sqrt{\kappa}(1-R)/(1-R\kappa)$ and $\kappa \equiv \exp(-\alpha L)$ provides the residual absorption of the intracavity EIT medium. As the effective cavity decay factor has gone down, the requisite cavity scan speed also goes down and is given by $v_{cav}^E = \frac{v_{cav}(1+\eta)}{[F^e]^2}$. Hence there is considerable reduction in cavity scan speed for a cavity containing EIT medium in comparison to an empty cavity provided $[F^e]^2 >> 1+\eta$. When the system reaches ideal EIT condition, then the following conditions can be satisfied, i.e., $\kappa = 1$ and $v_{cav}^E = v_{cav}/(1+\eta)$ [108]. The cavity enhancement factor F^e as a function of Ω_C is plotted in Fig.3.10. The value selected for the parameter γ_{31} is 0.1 MHz, which is quite close to the actual value for rubidium atomic vapor at room temperature. The cavity enhancement factor F^e or the cavity performance is greatly increased as Ω_C increases from zero to a value near 10 MHz. Beyond that value of Ω_C, further increase in Ω_C causes the cavity enhancement factor to reduce because of the presence of power broadening in the EIT peaks. The cavity enhancement factor F^e is also a sensitive function of γ_{31}. The sensitivity of F^e on γ_{31} is shown in the inset of Fig.3.10. Two cavity transmission profiles with respect to the normalized time t/T^E are shown in Fig.3.11. The dashed curve in Fig.3.11 represents the empty cavity transmission profile. Other parametric values used in the plot are $\Gamma_{cav} = 0.02$ leading to a cavity Finesse of 150 and $v_{cav} = 2 \times 10^4 \ \mu$m/s. Under these conditions the CRE can hardly be observed because the cavity Finesse is quite low and the cavity scan speed is also at the required lower limit. Note that the scan speed even at this lower limit is still quite large in view of what is available under practical conditions (for a sizable mirror mounted on a PZT). In Fig.3.11 the solid curve is depicting the cavity transmission profile with an EIT medium inside the optical cavity for $\Gamma_{cav} = 0.02$ and $v_{cav} = 4 \ \mu$m/s. Clear oscillation in the cavity transmission profile due to CRE appears under such setting. Without EIT medium, no oscillation can be seen under the same condition. The parametric conditions utilized in this numerical simulation are $\gamma_{31} = 0$ and $\Omega_C = 5$ MHz, which gives $\eta \simeq 24,000$. The above calculations show that the phenomenon of EIT can greatly help CRE from optical cavities with much lower finesses and with highly reduced cavity scan speeds. The curves in the inset of Fig.3.11 show the minimum scan speeds values to see CRE in the presence and absence of EIT medium as the cavity decay is varied. The other parameters used in the inset are the same as used in main Fig.3.11. It is easy to see that the cavity scan speed can be greatly reduced with the inclusion of an intracavity EIT medium (down to 4 orders of magnitude).

The methodology discussed here can be used to significantly enhance the

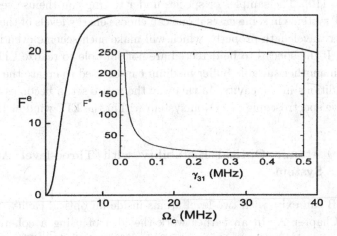

Fig. 3.10   Cavity enhancement factor versus coupling power for $\gamma_{31} = 0.1$ MHz. The inset shows the value of cavity enhancement factor as a function of γ_{31}. Other parameters are $R = 0.98$, $L = 5$ cm, and $l_0 = 36$ cm (adopted from [108] with permission).

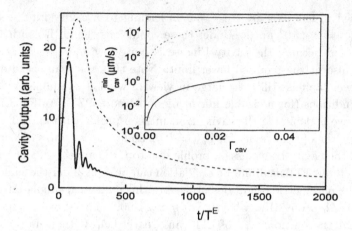

Fig. 3.11   Two examples of cavity transmission profile. The inset shows the required minimum cavity scan speed as a function of cavity decay. In these figures, the dashed curves are for an empty cavity and the solid curves are for the cavity with an intracavity EIT medium (adopted from [108] with permission).

usual CRDS for the quantitative detections of atomic and molecular samples or species with ultrahigh sensitivity. Although only one scheme of CRDS techniques has been used to show the effectiveness of EIT medium in enhancing the detection sensitivity and reducing the stringent requirements, the same trick can be applied to enhance other CRDS techniques [104-110]. The samples or species under testing can themselves be used as EIT systems in some cases if one can choose energy levels of the species and laser wavelength properly, which will make such scheme even more powerful. If the species to be detected are not suitable to realize EIT conditions then another suitable buffer medium can be used to create the needed EIT conditions in the cavity. In this way the entire setup becomes an ultrasensitive spectroscopic detection system within the EIT window [108].

3.2   Atomic Optical Bistability with Three-level Atomic System

AOB can exist with two-level atoms inside an optical cavity, as discussed in Chapter 2. In an earlier work the idea of using a coherently driven three-level atoms to observe AOB was discussed [38]. The proposal was based on using a non absorbing resonance arising due to trapped population in a coherent superposition of sublevels of the ground state. The most significant result of this work was the reduced threshold power to observe AOB in comparison to a two-level system. This work encouraged to study three-level systems experimentally [35]. The AOB from a coherently driven three-level atomic medium in Λ-type configuration inside a FP resonator was observed due to a nonlinear mechanism having its origin in transverse optical pumping and population trapping in ground-state sub-levels. Both dispersive and/or absorptive bistability were displayed by such system under Hanle resonance condition [35]. Since in the two-level system absorption and dispersion cannot be controlled effectively with experimental parameters, there is a lack of control to change threshold powers of both absorptive and dispersive AOB from this system. Hence the observation of AOB in three-level systems has distinct advantages over the two-level system. This is because in three-level systems absorption, dispersion, and nonlinear optical properties of the probe laser field are controllable due to the presence of another laser field (acting as a control field) interacting with nearby atomic transition and hence introducing induced atomic coherence (and producing EIT). Thus the controllability of linear and nonlinear optical properties of probe field around resonance allows to control the shape, width, threshold, and direction of the hysteresis curve in AOB from three-level systems. Such AOB behaviors in composite system of three-level atoms inside optical cavities were studied theoretically [36] in three-level systems in ladder-type and Λ-type configurations of their levels. Recently, a full controllability of AOB in its shape, width, threshold, and direction (rotation of hysteresis loop) have been experimentally demonstrated in a three-level atomic EIT system inside an optical ring cavity [4,5,42-46,117]. Some of these interesting studies are discussed in the following sections.

3.2.1 *Equations for three-level atomic dynamics and field propagation*

The atomic systems having three discrete energy levels can have ladder-type, Λ-type, and V-type configurations. In the following we will only consider a closed three-level Λ-type system. The system is shown in Fig.1.9(b), which has energy levels with energies E_i (i=1,2,3) ($E_2 > E_3 > E_1$). The atomic system interacts with the probe and coupling laser. The probe laser having frequency ω_P and amplitude E_P interacts with the dipole allowed transition $|1> \rightarrow |2>$ (having energy difference in terms of frequency given by ω_{21}) with a frequency detuning of $\Delta_P = \omega_P - \omega_{21}$. The coupling laser with frequency ω_C, and amplitude E_C couples to the other dipole allowed atomic transition $|2 > \rightarrow |3 >$ (having energy difference ω_{23}) with a frequency detuning of $\Delta_C = \omega_C - \omega_{23}$. The transition dipole-moment matrix elements for the two transitions are considered to be real just to keep the analysis simple. The Rabi frequencies of the probe and coupling fields, i.e., Ω_P and Ω_C, are considered to be complex quantities. The radiative decay rate from level $|2 >$ to level $|1 >$ ($|3 >$) is given by γ_{21} (γ_{23}) and the nonradiative decay rate between levels $|3 >$ to $|1 >$ is given by γ_{31}. The existence of induced atomic coherence between levels $|3 >$ and $|1 >$ by two-photon process is responsible for the observation of EIT phenomenon. The coupling field E_C is responsible for the creation of the coherence ρ_{31} in the steady-state. Under the semiclassical approach the Liouville equation of density operator in dipole and RWA is given by Eq.(1.58). The populations of this closed atomic system satisfy the relation $\rho_{11} + \rho_{22} + \rho_{33} = 1$. The linear and nonlinear optical properties of this coherently prepared medium can be obtained by solving equations of density-matrix elements in the steady-state, as described in Chapter 1.

AOB in a medium composed of homogeneously-broadened three-level atoms kept inside an optical cavity can be modeled similarly as done for the case of two-level AOB discussed in Chapter 2. To formulate the problem of AOB phenomena with this three-level EIT medium we again employ the standard model developed in Ref.[86]. This model, as discussed in Chapter 2, is comprised of a unidirectional ring cavity having four mirrors M_i ($i = 1, 4$), as shown in Fig.3.12 (which is very much similar to Fig.2.1 except an additional optical field E_C is injected into the EIT medium). Mirrors M_1 and M_2 have identical reflection and transmission coefficients R and T, respectively, satisfying the condition $R + T = 1$. Other mirrors M_3 and M_4 are considered as 100 % reflectors just to make analysis simple. The atomic sample of length L is kept in one of the arms of cavity, whose dynamics is governed by the density-operator equations (Eq.(1.58)). The combined electric field (that includes the probe and coupling fields) seen by the atoms can be written as

$$\vec{E} = (\vec{E_P}e^{-i\omega_P t} + \vec{E_C}e^{-i\omega_C t} + c.c.). \tag{3.16}$$

The probe field for the Λ-type EIT system at frequency ω_P, which interacts with the atomic transition $|1 >$ to $|2 >$, acts as the cavity field and circulates inside the cavity. The coupling field E_C applied to the transition $|3 >$ to $|2 >$, having frequency ω_C does not circulate inside the optical ring cavity. The two fields propagate through the atomic sample inside the cavity in the same direction, so the first-order Doppler effect in this three-level

EIT system can be eliminated [24]. In such case, the theoretical treatments for a homogeneously-broadened medium can reasonably reflect the real experimental system (an atomic vapor cell) used to demonstrate AOB and related effects. Comparing to the two-level AOB system, the coupling field in the current system behaves like a controlling field and will be termed control beam now onwards. The induced atomic polarization responsible for AOB is given by

$$P(\omega_P) = N\mu_{21}\rho_{12},\tag{3.17}$$

where N is number density of atoms and μ_{21} is the transition dipole-matrix element for the probe transition. The probe field E_P enters into the cavity from the partially transparent mirror M_1 and drives one of the atomic transitions in the EIT system as mentioned above. The control field manipulates the induced polarization $P(\omega_P)$ on the probe transition via induced atomic coherence effect. In this way the absorption, dispersion, and nonlinear properties of the atomic medium for the cavity field can be easily regulated [4]. For the optical ring cavity, the incident field E_P^I, the transmitted field E_P^T and fields at different physical locations along the length in the cavity ($E_P(0,t)$ and $E_P(L,t)$) obey certain boundary conditions [1,2] stated as follows (which are exactly the same as for the two-level AOB case, see Eq.(2.4) in Chapter 2):

$$E_P^T = \sqrt{T}E_P(L,t),$$
$$E_P(0,t) = \sqrt{T}E_P^I(t) + Re^{-i\delta_0}E_P(L,t-\Delta t),\tag{3.18}$$

where the length of the atomic sample is specified by the parameter L, and the time taken by light to travel from mirror M_2 to mirror M_1 via mirrors

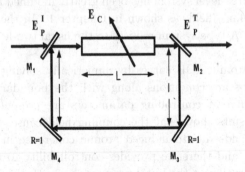

Fig. 3.12   A unidirectional ring cavity having four mirrors (M_1-M_4) and an atomic vapor cell of length L. M_3 and M_4 are perfectly reflecting ($R = 1$ for each). The incident and the transmitted fields are E_P^I and E_P^T, respectively, and the coupling field E_C is non-circulating in the cavity (reprinted from [4] with permission).

M_3 and M_4 is given by $\Delta t = (2l + L)/c$. The side arm between M_2 and M_3 as well as between M_1 and M_4 is of length 'l'. The phase detuning of cavity with respect to probe laser frequency is $\delta_0 = (\omega_{cav} - \omega_P)L_T/c$, where ω_{cav} is the frequency of nearest cavity mode to frequency ω_P and the total length of the ring cavity is $L_T \cong 2(l + L)$.

The equation providing dynamical evolution of the probe field inside the optical cavity is similar to Eq.(2.5) discussed in Chapter 2,

$$\frac{\partial E_P}{\partial t} + c\frac{\partial E_P}{\partial z} = 2i\pi\omega_P\mu_{12}P(\omega_P). \tag{3.19}$$

The cavity field obeys the boundary conditions as expressed in Eq.(3.18). The polarization $P(\omega_P)$ for the three-level atomic medium can be estimated by numerically solving the set of the density-matrix equations (Eq.(1.58)) in the steady-state limit, and then field can be obtained utilizing Eq.(3.18) to integrate Eq.(3.19) in the steady-state limit over the length of the sample. The boundary conditions in steady-state limit become (see Eq.(2.6) in Chapter 2)

$$E_P^T = \sqrt{T}E_P(L),$$
$$E_P(0) = \sqrt{T}E_P^I + Re^{-i\delta_0}E_P(L). \tag{3.20}$$

Although the field equation and the boundary conditions for the three-level atomic system are the same as the two-level ones, the medium polarization $P(\omega_P)$ for the three-level system has been greatly modified by the existence of control (coupling) field, as shown in Chapter 1. In the absence of the control field, the Λ-type system reduces to the usual two-level system [1,2].

The simulated atomic AOB curves by numerically solving Eqs.(1.58) and (3.19) in steady-state conditions along with the boundary conditions of Eq.(3.20) for different controlling parameters are presented in Fig.3.13. Clearly these results show that the coupling field causes the lowering of the AOB thresholds due to induced atomic coherence in the three-level atomic medium, and therefore provides controllability to the AOB curve [4,36]. The absorption at the line center reduces with the increase in control field strength as the EIT is enhanced. When the control (coupling) field is too large the bistability disappears due to the greatly reduced absorption and reduced nonlinear index at the line center in the EIT medium [33].

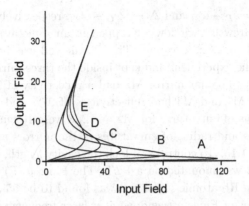

Fig. 3.13   The decrease in the thresholds due to the presence of the control (coupling) field for three-level atomic system in Λ−type configuration. Here $C = 400$, $\Delta_P = 1$, $\Delta_C = 0$, $\gamma_{21} = \gamma_{23} = 1$. Curves A, B, C, D, and E are for $\Omega_C = 1, 3, 5, 7$, and 10 respectively. All the parameters are in the unit of γ_{21} (reprinted from [4] with permission).

3.2.2   *Experiments on controlling the steady-state shape and thresholds of the atomic optical bistability*

The experiments discussed in the following are demonstration of AOB with three-level atoms in Λ-type configuration inside an optical ring cavity. In the experimental arrangement, both the probe and control beams undergo collinear propagation through the atomic medium to overcome the first-order Doppler effect [24]. The probe and control lasers are orthogonally polarized and combined by a polarization cube beam splitter (PB1) inside the cavity before entering the rubidium vapor cell, as shown in Fig.3.14. This experimental setup was similar to the one used for demonstration of cavity linewidth narrowing, except an electro-optical modulator (EOM) was placed in the path of probe field before it enters the cavity in order to modulate its intensity.

The AOB experiments were carried out in ^{87}Rb atomic vapor, using the D_1 line of the $5^2S_{1/2} \rightarrow 5^2P_{1/2}$ transitions as depicted in Fig.1.9(b). The circulating probe laser beam or the cavity field with frequency ω_P was tuned to near the atomic transition $|1> (F = 1, 5^2S_{1/2})$ to $|2> (F' = 2, 5\,^2P_{1/2})$ with frequency ω_{21}, while the control beam with frequency ω_c was tuned near the atomic transition $|3> (F = 2, 5^2S_{1/2})$ to $|2> (F' = 2, 5\,^2P_{1/2})$ with frequency ω_{23} . The frequency detunings of the probe and the control laser beams with respect to corresponding atomic transition frequencies are defined as $\Delta_P = \omega_P - \omega_{21}$ and $\Delta_C = \omega_C - \omega_{23}$, respectively. A 5-cm-long glass cell with Brewster windows, wrapped in an μ-metal sheet was used to confine rubidium atomic vapor. The atomic cell was heated to about 67.5 ^{0}C during the experiment and kept inside the three-mirror optical ring cavity (Fig.3.14). The flat mirror M2 had a reflectivity of 99%, while the concave mirrors M1 and M3 had reflectivities of 97% and 99.5%, respectively. The radius of curvatures for M2 and M3 were 10 cm [4,5,42]. Note that reflectivities and radius of curvatures of the mirrors are the same as described in Fig.1.15. To control and lock the cavity length, one of the concave mirrors M3 was mounted on a PZT. The Finesse (F) of optical ring cavity containing Rb atomic vapor cell was found to be 55 approximately. The measurement of Finesse was done at a laser frequency far away from any resonant absorption lines of ^{87}Rb. The total length of the optical cavity was 36.5 cm with a FSR of about 822 MHz. The entry of the probe laser beam was through mirror M1, which circulated in the optical cavity in a single direction. A polarization cube beam splitter (PB1) was used to send the control beam having orthogonal polarization with respect to the probe beam into the cavity, but it did not circulate in the cavity. In order

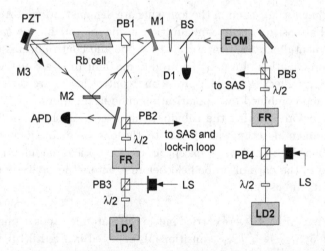

Fig. 3.14   The experimental arrangement used for studying optical bistability and optical multistability in ^{87}Rb atomic vapor: M1-M3 are mirrors of optical ring cavity, PZT is the piezoelectric transducer, LD1 and LD2 are control and probe lasers, respectively; PB1-PB5, polarizing cubic beam splitters; BS, beam splitter; EOM, electro-optic modulator; $\lambda/2$, half-wave plates; FR, Faraday Rotator; D1, detector; APD, avalanche photodiode detector; LS, Locking signal from reference Fabry-Perot cavity, SAS is a saturation atomic spectroscopy setup (adopted from [44] with permission).

to lock the frequency of the optical cavity to another FP cavity, one more diode laser (a third laser) was used as a reference beam. The three laser beams (probe, control, and reference) used in the experiment came from indigenously built extended-cavity diode lasers. The three diode lasers were frequency locked to their respective FP resonators dedicated for this purpose. The frequency detunings Δ_P and Δ_C of the probe and control lasers could easily be measured by using another FP cavity in association with a saturation absorption spectroscopy (SAS) setup. The intensity modulation of the probe laser beam entering the cavity was regulated by an EOM. The EOM was given a proper electrical signal from a function generator to produce a nearly triangular modulation for the intensity of the probe laser field (i.e., the cavity input field) during the experiment.

During the initial experiments of demonstrating AOB the temperature of the rubidium vapor cell was kept around 67-68 ^{0}C and the frequency of probe laser was adjusted near the desired transition of D_1 line of ^{87}Rb. The control laser frequency was also adjusted so that EIT could be seen. Next, the laser frequencies ω_P and ω_C were locked to certain predetermined detuning values with the help of their respective FP cavities dedicated for this purpose and scanning of the optical ring cavity was performed around its resonance by applying a ramp voltage on the PZT on which M3 was mounted [4,5,42]. The transmission profile obtained from the optical ring cavity was a symmetric one when the coupling laser beam was blocked from entering into the cavity. This situation is quite similar to the experimental condition when a two-level atomic medium was kept inside the optical cavity. Once the control beam was allowed to enter the cavity, the transmission profile became asymmetric, which can be attributed to the enhanced Kerr nonlinearity in the EIT medium due to generation of atomic coherence [33]. The degree of peak asymmetry was a sensitive function of the parameter Δ_C. After this initial setting for the experiment, the optical ring cavity's frequency was locked onto the transmission peak of the reference laser beam. Then the EOM in the passage of the probe laser beam (LD2), was turned-on, and a triangular modulation having a duration of 5 ms was applied to the optical field going in the cavity. When the control laser beam was absent in the optical cavity, no AOB could be found, but an optical-transistor-like behavior could be noticed in the input-output intensity plot. However, even in the presence of the control beam of quite high power, AOB still could not be observed if both $\Delta_C = 0$ and $\Delta_P = 0$. Setting one of the frequency detunings (Δ_P or Δ_C) nonzero, AOB started to appear for an extensive range of P_C [4,5,42]. Several experimentally observed AOB curves with different control beam parameters (such as Δ_C and P_C) are displayed in Fig.3.15, keeping other parameters, e.g., atomic density, cavity detuning, and Δ_P, unchanged. Figure 3.15(a) shows an observed AOB curve for the composite system of an optical cavity containing a three-level Λ-type atoms under the experimental conditions of $\Delta_P = 0$ MHz, $\Delta_C = 51.4$ MHz, $P_C = 1.5$ mW, and cavity detuning of 40 MHz approximately. Such AOB curve can be easily controlled by changing the control beam parameters. The value of Δ_C is changed to 25.7 MHz in Fig.3.15(b) with P_C fixed at 1.5 mW. From Figs.3.15 (a) and (b) it can be concluded that both the upper and lower switching thresholds (Y_1 and Y_2) and the width (Y_2-Y_1) of the AOB hysteresis curve could be controlled by simply changing the parameter Δ_C only [4,5,42]. The definitions of switching threshold intensities Y_1 (the lower one) and Y_2 (the upper one) are given as the locations in the AOB hysteresis curve where the quantity $dY/dX = 0$. In Figs.3.15(c) and 3.15(d), $P_C = 8.4$ mW was fixed, but Δ_C was tuned to 85.6 MHz and 256.8 MHz, respectively. By changing these control beam parameters, there are significantly noticeable changes in the shape of the AOB hysteresis curves. Hence it is possible to effectively control both the AOB range

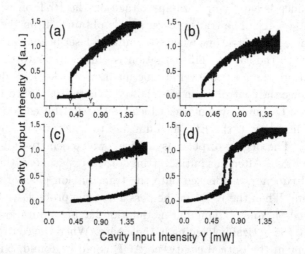

Fig. 3.15   The input-output intensity characteristics of the optical cavity containing three-level atomic medium with $\Delta_P = 0$. The other parameters are (a) $P_C = 1.5$ mW, $\Delta_C = 51.4$ MHz; (b) $P_C = 1.5$ mW, $\Delta_C = 25.7$ MHz; (c) $P_C = 8.4$ mW, $\Delta_C = 85.6$ MHz; and (d) $P_C = 8.4$ mW, $\Delta_C = 256.8$ MHz (reprinted from [42] with permission).

and switching intensity thresholds (Y_1 and Y_2) by manipulating the control laser parameters P_C and/or Δ_C. The values of threshold intensity in AOB hysteresis curves are very sensitive to the parameters such as the probe field frequency detuning Δ_P, cavity detuning, atomic number density or the cooperativity parameter C, control beam power P_C, and control field frequency detuning Δ_C. In two-level AOB case, when the frequency detuning of the laser beam is near or at zero, the AOB is caused by saturation and large nonlinear absorption. However, under the exact fulfillment of EIT (and/or CPT) conditions in the three-level system, i.e., when $\Delta_P = 0$, $\Delta_C = 0$ or $\Delta_P + \Delta_C = 0$, there is a great reduction in the absorption so the exact absorptive AOB could not be observed in this system. Once the parameters $\Delta_C \neq 0$ and $\Delta_P = 0$ (Figs.3.15(a) and 3.15(b)) are used to tune away from such EIT conditions in the experiment, the AOB behavior is observed and the switching thresholds come down when the control beam detuning gets smaller where absorption is larger outside the EIT window. Under the condition of $\Delta_C = 0$ and $\Delta_P = 0$, no AOB could be observed [4,5,42]. The appearance of AOB at low intracavity optical intensity and low control beam intensity is attributed to the enhanced Kerr nonlinearity under near resonance condition [33], which comes as a result of induced atomic coherence by the control beam interacting with the nearby transition under a little off-resonance condition, i.e., $\Delta_C \neq 0$. The AOB curves displayed in Fig.3.15(a) and 3.15(b) are typical absorptive type AOB, with intensity very low at the entire lower branch before reaching the threshold point.

The AOB exhibited by three-level EIT system can in general be a combination of absorptive and dispersive types. The dispersion phase detuning in such system is greatly enhanced by the Kerr nonlinearity in the EIT medium [33]. By increasing Δ_C to a larger value, the Kerr nonlinearity in the system reduces down, which leads to the increase in the switching intensity Y_2, as well as the width of the hysteresis cycle as shown in Fig.3.15(c). When the control beam detuning gets very large, e.g., $\Delta_C = 256.8$ MHz, as shown in Fig.3.15(d), the available nonlinear refraction is only due to the frequency detuning. So, the behaviors of AOB, as shown in Figs.3.15(c) and 3.15(d) are quite similar to the two-level dispersive AOB situation [42]. It is interesting to notice that in two-level AOB, the dispersive AOB is obtained by large frequency detuning Δ_P since it is the only field interacting with the atoms. However, in the three-level system, the AOB can be changed from absorptive (Figs.3.15 (a) and (b)) to dispersive (Figs.3.15(c) and 3.15(d)) type by increasing the frequency detuning Δ_C of the additional control beam. Therefore the shape and width of AOB curves, and even the types of the AOB, are very effectively controlled in the three-level EIT medium inside the optical cavity by the additional control beam. The observed experimental results are very much in good agreement with the earlier work related to the measurements of nonlinear refractive index in this system [33]. The cooperativity parameter C (defined after Eq.(2.12)) can be obtained with the help of the measured atomic absorption and cavity Finesse when the system displays EIT behavior in the experiment. The important point to note here is the advantage of AOB controls in three-level atomic system with experimental parameters which are not so easily available in the two-level atomic media. The external controlling parameters for the three-level EIT system (other than the atomic density) are the frequency detuning Δ_C and power P_C of the control laser beam. To quantify this statement, the switching threshold intensity ratio Y_2/Y_1 is plotted as a function of Δ_C in Fig.3.16, for two different values of the control laser intensities. In this figure, curves A and B are for $P_C = 8.4$ mW and $P_C = 1.5$ mW, respectively [42]. For low values of Δ_C, the ratio Y_2/Y_1 is also small, implying a lower AOB threshold due to the enhanced near resonance Kerr nonlinearity. As Δ_C is increased, Y_2/Y_1 also increases because the Kerr nonlinearity is reduced down considerably at large detuning [33]. When Δ_C gets very large, Y_2/Y_1 is again reduced, which is due to the effect of dispersion caused by Δ_C alone. At quite large Δ_C, the Kerr

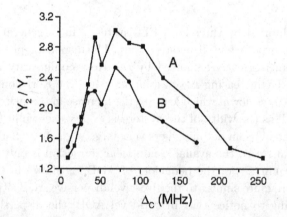

Fig. 3.16   Ratio of switching threshold intensities Y_2/Y_1 as a function of Δ_C. Curve A is for $P_C = 8.4$ mW and curve B is for $P_C = 1.5$ mW (reprinted from Ref.[42] with permission).

nonlinearity reduces by a significant amount and the system starts behaving like a two-level atomic system under such condition. The variation of switching thresholds (for a fixed Δ_C) as a function of P_C is also studied for this system and results are plotted in Fig.3.17. Curves A and B represent switching threshold ratio Y_2/Y_1 under the condition of $\Delta_C = 42.8$ MHz and 17.1 MHz, respectively. There is a rising trend in Curve A with increasing P_C. This is due to the reduction in the Kerr nonlinearity with increasing control power P_C for this particular value of the parameter Δ_C. On the contrary, an opposite behavior in curve B is observed, i.e., the switching ratio slightly going down with the increase of P_C. This can be explained as an increase in nonlinearity with the increase of P_C for this value of Δ_C [33,42].

The AOB behaviors so observed with the coherently-prepared three-level EIT medium inside an optical cavity are very much different from the earlier investigations of AOB from two-level atomic media [1,2], as discussed in Chapter 2. The induced atomic coherence near EIT resonance causes great changes in the linear and nonlinear properties of the three-level atomic medium for the probe field [24,33]. The benefits of utilizing such three-level atomic medium inside an optical cavity is the simplicity of experimental operation and controllability with experimental parameters. In two-level atomic systems, one has to use either atomic beams [87,88], buffer gas [2], or cold atoms [17] to suppress the large Doppler broadening, so the bistability phenomenon could be observed. In most of these cases, the experiments

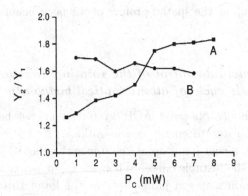

Fig. 3.17   Ratio of switching threshold intensities Y_2/Y_1 as a function of P_C. Curve A is for $\Delta_C = 42.8$ MHz and curve B is for $\Delta_C = 17.1$ MHz (reprinted from Ref.[42] with permission).

of AOB in two-level systems had to be done using vacuum chambers. However, by collinear propagation of the control and probe fields through the atomic vapor cell filled with the three-level Λ-type rubidium atoms, the first-order Doppler effect can be eliminated [24] and the experiment can be done without using vacuum systems, which has greatly simplified the experiment [4,5,42].

In order to understand the basic physical phenomena and practical applications of AOB, some optimal conditions for desired AOB are required. The experimental demonstrations of controlled AOB in the three-level atomic system with the control beam intensity and frequency detuning can certainly be viewed as a remarkable step towards achieving this aim. By making use of the knowledge developed in those studies, it will be easy to design more versatile all-optical switches and logic gate devices for optical computing and quantum information processing with controllable three-level EIT systems, most likely in solid-state materials. In this demonstrated intracavity EIT system, there are reduced absorption and enhanced Kerr nonlinearity via atomic coherence, hence the switching threshold values are substantially reduced to allow one to make highly efficient all-optical switches operating at much weak light intensity levels. The simplified theoretical model, described in subsection 3.2.1, has provided qualitative understanding of the three-level AOB and its basic behaviors. To completely understand these interesting AOB phenomena will require a more detailed theoretical calculations to match the detail experimental conditions, including multi-Zeeman sublevels and various Raman processes near the EIT resonances, as well as the spatial profiles of the laser beams during propagation [5].

3.2.3   *Experimental control of the rotating direction of the hysteresis cycle of atomic optical bistability*

The most commonly observed AOB hysteresis curves have a counterclockwise rotation due to energy considerations. As the cavity input intensity increases, it will jump up at the upper threshold Y_2 and go to the upper branch of the bistable curve. When the input power is lowered from the far upper branch, it will jump down at the lower threshold value Y_1 back to the lower branch of the bistable curve. Such hysteresis loop is called 'forward' hysteresis cycle (HC) (or forward HC) of the bistability. In AOB with two-level atomic medium, the observed counter-clockwise (or forward) HC is caused by the dissipation in the system [2]. Most other smart materials, such as ferroelectric and ferromagnetic (or ferroic) materials, also show forward HCs [118]. In an earlier reported optical bistability experiment using a semiconductor medium, consisting of a 4.2 μm thick GaAs etalon with working temperature of 80 K, the competition between electronic nonlinearity due to free-exciton and the thermal effect has made the switching down intensity to be higher than the switching up intensity in the optical bistability curve, which resulted in a HC going with a clockwise rotation called 'backward' hysteresis cycle (HC) (or backward HC) [119]. A prominent role of the cavity detuning on the HC was also observed in that experiment. A cavityless optical bistability system consisting of a CdS semiconductor medium also displayed backward HC [120]. In that system the optically induced absorption change due to thermal effect near the free and bounded excitons played a crucial role [120].

The three-level EIT system discussed above for controlling AOB is very capable and flexible with respect to changes in experimental parameters. Using that system it was further demonstrated [44] that the direction and width of the HC in AOB could be efficiently controlled with the variation of the control laser frequency detuning (Δ_C) alone, holding other experimental

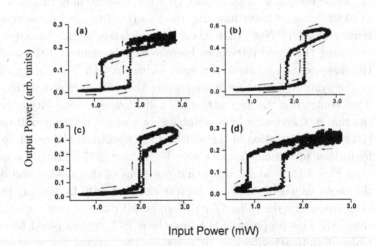

Fig. 3.18   The observed input-output intensity characteristics of the system. The parameters are $\Delta_P = 0$, $\Delta_\theta = 50$ MHz, $P_C = 14.0$ mW, $T = 65\ ^0C$, and (a) $\Delta_C = 103$ MHz; (b) $\Delta_C = 137$ MHz; (c) $\Delta_C = 171$ MHz; (d) $\Delta_C = 275$ MHz. As clearly seen, the HC changes shape and rotation direction as the control beam detuning increases (adopted from [44] with permission).

parameters unchanged. The results of this experimental demonstration are depicted in Fig.3.18. Figure 3.18(a) displays the AOB hysteresis loop with parameters $P_C = 14.0$ mW, $T = 65$ ^{0}C, Δ_θ (the cavity detuning)$= 50$ MHz, $\Delta_C = 103$ MHz, and $\Delta_P = 0$, which is a usual forward HC if one considers the main hysteresis loop. This is because in this HC, the up-switching threshold intensity is larger compared to the down-switching threshold intensity. Such intensity changing directions can be directly observed on the X-Y oscilloscope used to record the input-output intensity curves, as shown in Fig.3.18. By increasing the value of Δ_C to 137 MHz, the HC has modified significantly and resembles to an arabic-roman numeral figure 'eight' (Fig.3.18(b)) [4,44]. The arrows marked on the HC show the path taken by the cavity field intensity when the input intensity is scanned first to a higher value and then down to a lower value using the EOM. It is straightforward to notice both the usual forward HC, i.e., the lower part of the hysteresis curve as well as the backward HC, i.e., the upper part of the hysteresis curve in this figure. When the value of Δ_C is further increased to 171 MHz, the system exhibits an extraordinary change in the width and the direction of the main part of the hysteresis curve as shown in Fig.3.18(c). There is change in the width of the main HC, which has been significantly decreased and now this HC moves in the backward or clockwise direction. This implies that the up-switching threshold intensity of AOB becomes lower in comparison to the down-switching threshold intensity of AOB. Note that when Δ_C reaches a particular value, the two switching threshold intensities become exactly same and the main part of HC disappears, i.e., there is a zero width for HC. The energy dissipation in the system can be estimated using the area enclosed within the HC. Thus controlling the area of the HC will allow controllability of energy dissipation in the system by a suitable choice of the experimental parameters [4,44]. The elongation in the width of the backward HC could be achieved by further increasing Δ_C to a very high value (275 MHz). It is quite clear from Figs.3.18(a) and 3.18(d) that positions of the upward and downward threshold values of AOB can be interchanged only by varying the control laser frequency detuning (Δ_C), keeping all other system parameters unchanged. This property of the three-level EIT system could be utilized to make all-optical switchings by reversing the control beam detuning (Δ_C) between two values, which will allow switching of the cavity intensity between upper and lower branches. The variation of other parameters such as Δ_θ can also give similar effect of going from forward to backward HC in the this system [4,44].

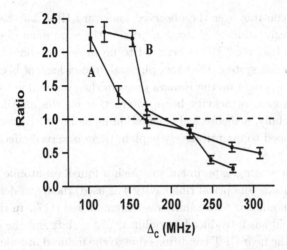

Fig. 3.19 Ratio of upper to lower switching threshold intensities as a function of Δ_C with $\Delta_P = 0$, $\Delta_\theta = 50$ MHz, and $P_C = 14.0$ mW. Curves A (■) and B (●) are for $T = 65\ ^0C$ and $T = 70\ ^0C$, respectively (reprinted from [44] with permission).

The characteristics of this system was further investigated in terms of the ratio of upper switching threshold intensity to lower switching threshold intensity of the AOB, as displayed in Fig.3.19 for two different temperature conditions of the system [4,44]. The temperature of the sample cell in curves A and B were 65 ^{0}C and 70 ^{0}C, respectively. The ratio of the threshold intensities decreases as Δ_C is increased. The HC reverses its direction when the ratio crosses the magnitude of 1. The trend in the change of ratio with Δ_C at two different temperatures is very much the same except the ratio is toward the higher side at higher temperature. The shape change and transition from forward HC to backward HC occurs (i.e., the ratio crosses the magnitude 1) when the Δ_C crosses a certain value at a given temperature. The physical mechanism of the backward HC observed in this three-level EIT system must be very different from those reported in earlier works [119,120]. The main difference arises due to the fact that thermal effect (the mechanism attributed to the backward HC in the earlier works [119,120]) could not contribute to the atomic system dynamics because the thermal effect is a very slow process and could not exist in the atomic vapor. The system discussed here will have more advantages in potential applications for all-optical switching and all-optical storage devices because its dynamics is fast. Also, in the previous experiments, there were lack of controls over the observed backward HCs, but here the control can be easily achieved through adjusting the experimental parameters. Although the backward HC is believed to be caused by some kind of gain mechanism in this system, the exact physical process has not been identified theoretically. It could be the Raman gains in the system due to the strong control beam and intracavity beam interacting on the multilevel atomic system. Hopefully, a theoretical calculation including all Raman processes can be developed to quantitatively explain these observed effects [44] .

In another interesting experiment in which a four-level atomic medium of ^{87}Rb atoms inside an optical ring cavity was used, the dependence of AOB curve on the nonlinear phase shift was demonstrated [117]. In that system, the control field has introduced a nonlinear phase shift and the atomic system satisfies the near CPT conditions due to the induced atomic coherence. The root of this observed effect lies in the absorptive and refractive nonlinearities of the atomic medium enhanced by atomic coherence. This effect has been utilized to demonstrate an all-optical flip-flop and storage of optical signals with a low peak power of several tens of microwatts [4].

Chapter 4

Atomic Optical Multistability

4.1 Atomic Optical Multistability: Introduction

The phenomenon of optical multistability (OM) is yet another common property of nonlinear optical systems with feedback and it means that there are more than two stable output states of a system corresponding to different amplitudes or polarizations of the optical field for a given input field intensity to the system. Using a simplified setup of a FP cavity containing atoms with many degenerate or nearly degenerate sub-levels, e.g., Zeeman sublevels in ground state, undergoing optical pumping, optical tristability (multistability) was predicted in eighties [121-124]. For a linearly-polarized incident light, three stable states could be observed in the polarization of the transmitted light leading to multiple HC and symmetry breaking bifurcation ascribed to competition between hyperfine and Zeeman pumping processes. In such early experiments magnetic field and high pressure buffer gases were used, and the Zeeman coherence was found to be responsible as an important mechanism for obtaining OB and OM in these works [121-124]. This kind of optical tristability was also discussed theoretically in the context of a butterfly catastrophe in a three-level system in Λ-type configuration under a large atomic detuning with no saturation [125] effect and the nonlinearity of system was found to be responsible for this kind of multistable behavior. This phenomenon was then experimentally observed in sodium atomic vapor system [126]. The theoretical work of Ref.[125] was further generalized to include saturation in the dispersive limit which made the steady-state of system unstable and thus facilitated period-doubling, optical turbulence etc [39]. By including the effect of ground-state coherence both optical tristability as well as higher-order bistability in sodium atomic vapor system under near resonant D_1 excitation were predicted

115

[127]. Note that atomic optical multistability (AOM) is a property of the composite atom-optical cavity systems and basically it refers to systems which are neither stable nor totally unstable, but the system output alternates among more than two mutually exclusive states over time.

The recent progress made towards the controllability of observed AOM in the three-level atomic system inside an optical ring cavity will be discussed in the following [4,43]. Due to exact knowledge and controllability of the linear absorption, dispersion, and Kerr nonlinearity at various control field and cavity field parameters, e.g., optical powers and frequency detunings of these fields, as discussed in Chapter 1, one can control and manipulate the shape and direction of rotation of AOM in such composite system by simply changing various experimentally controllable parameters. The observed AOM has been explained with the help of simultaneous presence of the absorptive and dispersive types of AOBs, and their combined hysteresis behaviors [43], in a bit different manner, from the previously given mechanisms.[39] for AOM.

4.2   Atomic Optical Multistability with Three-level Atoms inside an Optical Cavity

4.2.1   *Theoretical calculation*

The theoretical model describing AOM in the three-level atomic system is exactly the same as what has been used in subsection 3.2.1 for the AOB in N (number density) homogeneously-broadened three-level atoms confined in an optical ring cavity. The dynamics of the atomic system is governed by the density-matrix equations (Eq.(1.58)). The electric field seen by the atoms, which are in Λ-type configuration (Fig.1.9(b)), is composed of the coupling (control) field E_C at frequency ω_C applied to the transition $|3>$ to $|2>$ (which does not circulate inside the optical ring cavity). The probe field for the EIT system at frequency ω_P (interacting with atomic transition $|1>$ to $|2>$ and satisfying the boundary conditions of Eq.(3.18)) circulates inside the cavity as the cavity field. The coupling field in the EIT system acts as a mere control field. The composite field seen by the atoms is given by Eq.(3.16) and the induced atomic polarization on the probe transition responsible for observing AOM is given by Eq.(3.17) (see in Chapter 3) together with the solved density-matrix element ρ_{12} from Eq.(1.58). In the absence of the control field, the Λ-type system reduces to the usual two-level system [1,2].

The expression of induced susceptibility in the two-level atomic system depends on the input field quadratically as [1]

$$\chi = \alpha(1 - i\Delta)[1 + \Delta^2 + |E_P|^2/I_s]^{-1}, \tag{4.1}$$

which is also the Eq.(2.9) in Chapter 2. The two-level system shows AOB under normal circumstances of homogeneous broadening in the system, but it can also exhibit AOM behavior under certain parametric conditions as we show in the following. For this purpose we consider E_P as a complex quantity and define $E_P = \varepsilon_P(z)\exp(i\Phi_P(z))$, which leads to the following equations for the amplitude $\varepsilon_P(z)$ and phase $\Phi_P(z)$ using the Maxwell's equation in the steady-state [1]

$$\frac{d\varepsilon_P}{dz} = -\chi_a(\varepsilon_P^2)\varepsilon_P,$$

$$\frac{d\Phi_P}{dz} = -\chi_d(\varepsilon_P^2). \tag{4.2}$$

The quantities χ_a and χ_d are the absorptive and dispersive parts of susceptibility χ, given by Eq.(2.8) of Chapter 2. By defining the quantities

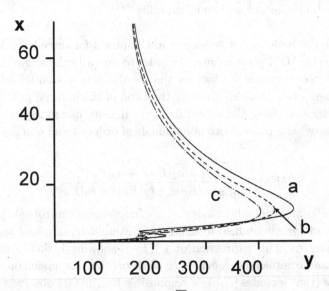

Fig. 4.1 Optical multistability with $x = \sqrt{X}$ as the normalized transmitted field. In all curves $C = 900$, $\Delta = 5$, $\theta = 0.05$. Curve **a** corresponds to the case of homogeneously broadening ($T_2^* = \infty$). In **b**, $\gamma_\perp T_2^* = 1$; in **c**, $\gamma_\perp T_2^* = 0.5$. Clearly, multistability disappears (in **b** and **c**) with the increase of inhomogeneous broadening (adopted from Fig.6 of [1] with permission.)

$Y = \frac{P_I}{I_s T}$ and $X = \frac{P_T}{I_s T}$, where $P_I = (E_P^I)^2$ and $P_T = |E_P^T|^2$ are incident and transmitted intensities, respectively [1], the expression for the cavity transmissivity is given by using the boundary conditions (Eq.(2.6) of Chapter 2) as [1]

$$\Upsilon = \frac{X}{Y} = \frac{T^2}{(\zeta(X) - R)^2 + 4R\zeta(X)\sin^2\{\frac{1}{2}[\Delta\ln(\zeta(X)) - \delta_0]\}}. \quad (4.3)$$

Also, we have used the fact: $\zeta = \frac{\varepsilon_P(0)}{\varepsilon_P(L)}$, and $\Phi_P(L) - \Phi_P(0) = \Delta\ln(\zeta(X))$ [1]. Note that the cavity transmissivity function Υ is a sensitive function of $\zeta(X)$. The system can exhibit multistability (see Fig.4.1) depending on the number of oscillations that the function $\sin\{\frac{1}{2}[\Delta\ln(\zeta(X)) - \delta_0]\}$ makes, and such oscillations of the sine function are governed by the quantity $\alpha L \Delta/(1 + \Delta^2)$. The quantity η takes values from 1 to $\exp[\alpha L/(1 + \Delta^2)]$ and consequently $\Delta\ln\zeta - \delta_0$ goes from $-\delta_0$ to $\alpha L\Delta/(1 + \Delta^2) - \delta_0$ [1]. However, in the mean field limit the above Eq.(4.3) reduces to Eq.(2.19) which does not show any multistability. The inhomogeneous broadening appears in the expression through the parameter $T_2^*\gamma_\perp(= T_2^*\gamma_{21})$ and its effect on multistability is shown in Fig.(4.1). For increased inhomogeneous broadening, the system does not show multistability [1].

We can use the basic model developed in Chapter 3 for three-level AOB to investigate the AOM phenomenon. In order to quantify the origin of AOM in this composite system we look at the steady-state susceptibility χ^Λ in the presence of control field, given by the ratio of the induced polarization $P(E_P)$ to incident field E_P, solved from the density-matrix equations. In general, the χ^Λ is a ratio of two polynomials of orders 4 and 6 in E_P in the form [36]

$$\chi^\Lambda(E_P) = \frac{[a_1 + a_2|E_P|^2 + a_3|E_P|^4]}{b_1 + b_2|E_P|^2 + b_3|E_P|^4 + b_4|E_P|^6}, \quad (4.4)$$

in which the complex numbers a_i and b_i are functions of control field parameters Δ_C and P_C, probe field detuning Δ_P, atomic density, and radiative decay constants. This expression of $\chi^\Lambda(E_P)$, showing higher-order nonlinearities and complicated dependence of absorption/dispersion on system parameters (that includes $|E_P|$), is responsible for the OM observed in this three-level atomic system inside an optical cavity. In the case of two-level atoms $\chi^\Lambda(E_P)$ reduces to a ratio of zero-order to quadratic polynomial in E_P (Eqs.(2.9) and (4.1)) and hence, can only show bistable behavior as described in sections 2.3 to 2.5. Since there is no control field in two-level

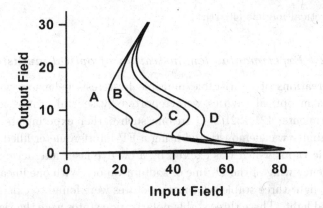

Fig. 4.2 The optical multistability due to the presence of the control field for three-level atomic system in Λ−type configuration. Curves are generated by solving Eq.(1.58) in steady-state together with Eq.(3.19) using parameters $C = 200$, $\Delta_P = -4.5$, $\Delta_C = 4.5$, $\gamma_{21} = \gamma_{23} = 1$. Curves A, B, C, and D are for $\Omega_C = 2.0$, 4.0, 6.0, and 7.0, respectively. All the parameters are in the unit of γ_{21}.

atoms so a_i and b_i do not contain parameters of control field. The numerical simulations of controlled multistability for a three-level system in the Λ-type configuration of its levels are shown in Fig.4.2.

The total linear and nonlinear susceptibility for a collection of three-level atoms (having number density N) in Λ-type configuration under weak field approximation for E_P is given by [33,34]

$$\chi_{app}^{\Lambda} \cong \frac{iN|\mu_{21}|^2}{\hbar} \frac{1}{F} [1 - \frac{2\gamma_{31}}{2\gamma + \gamma_{21}} - \frac{|\Omega_P|^2}{2\gamma + \gamma_{21}} \frac{F + F^*}{|F|^2}], \qquad (4.5)$$

with $F \equiv \gamma - i\Delta_P + (|\Omega_C|^2/4)/[\gamma_{31} - i(\Delta_P - \Delta_C)]$. Note that Eq.(4.5) is similar to Eq.(1.61), where some more details are given. The first term in the expression of Eq.(4.5) is the linear susceptibility, the second term is the contribution to the linear susceptibility from the higher-order density-matrix element, and the last term is the third-order (or Kerr-like) nonlinearity (see Eq.(1.62) also) due to the finite probe intensity modified by the atomic coherence. One can rewrite Eq.(4.5) as $\chi_{app}^{\Lambda} \cong \chi^{(1)} + 3\chi^{(3)}|E_P|^2$. The expression in Eq.(4.4) is more general than the one in Eq.(4.5), and contains higher order nonlinearities responsible for multistabilities.

Controlling OB to OM behavior using spontaneously generated coherence (arising due the interference in the nearby decaying atomic levels) in there-level atomic media inside an optical ring cavity has been predicted using theoretical models [40,41].

4.2.2 *Experimental demonstration of optical multistability*

Observations of multistable/multiple hysteresis behaviors with nonlinear media in optical cavities were reported by several groups in some early experiments [121,122,126]. In one such earlier experiments [126] optical tristability was demonstrated using a FP interferometer filled with sodium atomic vapor which was excited by a cw dye laser interacting on the high frequency wing of the D_1 line of sodium vapor. With one linearly polarized input light three stable polarization states were found to exist in the transmitted light. These three stable polarization states were the right-circularly polarized state σ^+, left-circularly polarized state σ^- and linearly polarized state. The two circular polarization states were found to be the dominant ones (see Fig.4.3). These results were theoretically explained with the model given in Ref.[125], where spin polarizations induced by optical pumping of degenerate Zeeman levels in the ground state of the sodium atoms under anomalous dispersion condition were responsible to give this phenomenon. The multistability experiment reported in Ref.[121] has shown

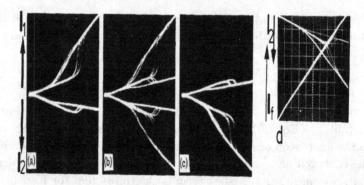

Fig. 4.3 Optical tristability (output vs input intensities) as monitored on the scope (see Fig.2 of Ref.[125] for comparison with theoretical prediction) showing hysteresis for both circular components of the transmitted light beam. 1.8 GHz detuning, 8 mbar Ar. As for all the following pictures, I_1 and I_2 indicate the two circularly polarized output signals. (a) Input-polarization-controlling Pockels-cell bias voltage (PCV)-1 V with respect to (b); hysteresis observed only from linear to σ^+ dominant state. (b) symmetrical condition. (c) PCV+1 V with respect to (b); hysteresis observed only from linear to σ^- polarization dominant state. (d) Fluorescence viewed orthogonally to the beam direction (I_f, bottom trace); also transmission for one circular component (I_2, top trace)(adopted from Fig.2 of [126] with permission.)

interesting hysteresis loops in the transmission profiles of a sodium vapor filled FP resonator. No buffer gas was used in that experiment and the laser entering the FP resonator was tuned near the D_1 line of sodium atom under different conditions of temperature and magnetic field. The cavity transmission showed multiple hysteresis cycles and symmetry breaking bifurcations which were physically interpreted in terms of the competition between hyperfine and Zeeman pumping processes because there were two counter-propagating optical fields interacting with the atomic medium. These previous experiments demonstrating multistability behaviors have all involved complicated multi-Zeeman sublevels in the atomic media, which would be hard to analyze.

The recent experiment of demonstrating AOM in the system of three-level atoms inside an optical ring cavity [43] is completely different from the previously studied multistability behaviors in atomic systems [121,122,126]. In this new experiment, the atomic medium could be considered as having a simple three-level Λ-type configuration (as shown in Fig.1.9(b)). In such three-level EIT medium, the induced atomic coherence causes significant alteration to the absorption, dispersion, and nonlinearity for the cavity field [33], as demonstrated for the controlled AOB discussed in Chapter 3. Since the experimental parameters are easily adjusted, AOM phenomenon has shown to exist in certain parametric regions [43]. It was demonstrated

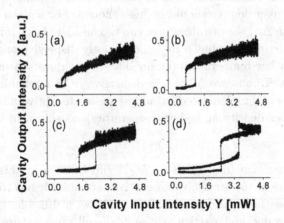

Fig. 4.4 The input-output intensity characteristics of the system for different atomic number densities. The parametric conditions are $\Delta_P = 0$, $\Delta_C = 42.8$ MHz, $P_C = 8.5$ mW, and (a) $T = 65.4$ ^{0}C; (b) $T = 74.5$ ^{0}C; (c) $T = 85.3$ ^{0}C; (d) $T = 92.5$ ^{0}C (reprinted from [43] with permission).

that as the control field frequency detuning (introducing dispersion for the cavity field) and/or the temperature of the vapor cell (which is directly related to the atomic absorption) are increased further from the conditions discussed in Chapter 3 for controlling AOB, multistable hysteresis scenerio starts appearing [43]. The simultaneous presence of the mixed absorptive and dispersive AOB mechanisms play the essential roles in generating the multistability behaviors in this system. For example, the augmentation of atomic number density brings the dispersive-type AOB into multistability with added absorption for the cavity field [43]. On the other hand, the increased absorption for the probe beam at near resonance brings absorptive AOB in prominence, however the additional frequency detuning (Δ_C) for the control beam, introduces an additional dispersion into the system. So, it is the control field frequency detuning Δ_C, not the cavity (probe) field frequency detuning Δ_P, that switches the pure absorptive-type AOB to multistability with both absorptive and dispersive characteristics [43]. In the following we discuss the experimentally measured results in details.

Figure 4.4 shows the controlled transition from AOB to AOM by going up in the atomic number density via increasing the temperature of Rb vapor cell [43]. The parameters set for this experimental measurements were $\Delta_P = 0$, $\Delta_C = 42.8$ MHz, and $P_C = 8.5$ mW, and curves (a), (b), (c), and (d) are for $T = 65.4\ ^0$C, $T = 74.5\ ^0$C, $T = 85.3\ ^0$C, and $T = 92.5\ ^0$C, respectively. It can be concluded that in order to observe multistablity behavior, atomic number density should be high enough. For a given dispersion due to the finite Δ_C, the bistable curve can be considered as a mixed absorptive plus dispersive AOB, but a more dominantly dispersive one, especially at a relatively low temperature (or small absorption), as shown in Figs.4.4(a) and 4.4(b). To observe AOM, the dispersive AOB mechanism needs to be appropriately complemented by adding more absorptive mechanism (from high number density at higher temperature), as shown in Figs.4.4(c) and 4.4(d).

Similar transition from AOB to AOM can be seen in this system when temperature of the atomic vapor cell is kept constant but the control laser frequency detuning Δ_C is increased, as shown in Fig.4.5. For a relatively small Δ_C value and particularly at high cell temperature, the intensity input-output curve of the cavity shows a typical absorptive AOB hysteresis curve (Fig.4.5(a)), however, as Δ_C increases, but keeping all other parameters unchanged, clear AOM behavior starts emerging out [43]. In this sit-

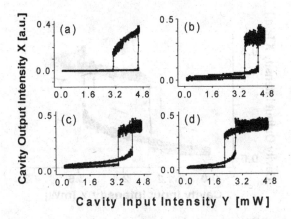

Fig. 4.5 The input-output intensity characteristics of the system for different frequency detunings of the coupling field. The parametric conditions are $\Delta_P = 0$, $T = 91.5$ ^{0}C, $P_C = 8.5$ mW, and (a) $\Delta_C = 17.1$ MHz, (b) $\Delta_C = 25.7$ MHz, (c) $\Delta_C = 51.4$ MHz, and (d) $\Delta_C = 214.0$ MHz (reprinted from [43] with permission).

uation for a given high number density, meaning a fixed high temperature of vapor cell, the absorptive AOB needs to be appropriately complemented by the large dispersion (as provided by increasing Δ_C) in order to observe such AOM hysteresis, as displayed in Figs.4.5(c) and 4.5(d).

4.3   Mixed Absorptive and Dispersive Effects in Atomic Optical Multistability

At the time of measuring AOM in the experiments the experimental parameters were varied continuously so that appropriate parametric regions could be explored to observe such AOM behaviors. From Figs.4.4 and 4.5, it is clear that AOM could only be observed with certain parametric regions where both absorptive and dispersive type AOB can coexist. One of the typical AOM curve for certain absorptive and dispersive parametric regime is displayed in Fig.4.6 with the experimental conditions of parameters: $P_C = 8.5$ mW, $T = 90.5$ ^0C, $\Delta_C = 43$ MHz, and $\Delta_P = 64$ MHz [43]. This experimental curve clearly shows more than two stable points at certain input intensity. In fact, it is easy to find that there are a pair of three stable steady-state points for certain given input intensity values on the extended hysteresis curve of Fig.4.6. It is quite straightforward to understand the input-output intensity characteristics by means of the arrows drawn on the curve. The paths taken by the cavity field intensity, as the in-

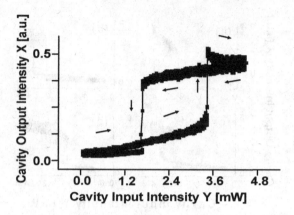

Fig. 4.6 Observed OM in the input-output intensity characteristics of the system with the parameters $\Delta_P = 64$ MHz, $\Delta_C = 43$ MHz, $T = 90.5\ ^\circ$C, and $P_C = 8.5$ mW (reprinted from [43] with permission).

put field strength increases from zero to about 4.5 mW (measured in terms of power) and back down to zero value again, are marked by arrows, which are established by observing x-y plot on the oscilloscope. These paths indicated by the arrows suggest that this hysteresis curve is not the usual AOB curve but more complicated one. In order to better understand the mechanism for forming this kind of hysteresis curve, two calculated curves, (i) and (ii), which represent the pure dispersive and pure absorptive bistability plots (using equations discussed in subsection 3.2.1), respectively, are drawn in Fig.4.7 for the parameters close to the experimental conditions used in Fig.4.6. When curve (i) is plotted, the absorption contribution from the input-output intensity expression has been removed deliberately. Similarly, for plotting curve (ii), the dispersion contribution from the input-output intensity expression has been taken out deliberately. Therefore, curves (i) and (ii) show the dispersive only and absorptive only AOB characteristics, respectively. As pointed out by the arrows on the two curves, when the cavity input intensity increases from zero, the cavity output intensity goes along the dispersive bistable curve (i) to its upper threshold point A, where the output intensity jumps up to the upper branch of the dispersive curve at point B, and further goes along this dispersive bistable curve (i) on the upper branch to the high value [43]. However, when the input intensity is scanned downward from the high value, the output intensity can choose to switch to the upper branch of the absorptive bistable curve (ii) at the crossing point E and goes along it down to its lower threshold point C, where

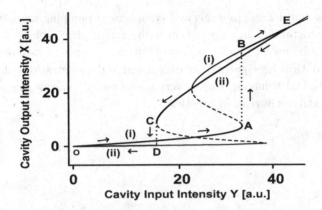

Fig. 4.7 Construction of optical multistable behavior (represented by the dark lines along with arrows) with the help of a pure-dispersive (curve (i)) and a pure-absorptive (curve(ii)) optical bistability curves co-existed in the system with parameters similar to the experimental parameters. (reprinted from [43] with permission).

it jumps down to point D on the lower branch, then goes back to zero. By looking at the entire path (the solid curve OABECDO) taken by the cavity output intensity, it really shows the similarity to the multistability obtained in the experiment, as shown in Fig.4.6. The new hysteresis cycle (solid line) is confined on the stable parts belonging to the two bistable curves (i) and (ii). From such comparison, one can explain the mechanism of the experimentally observed AOM to be caused by the simultaneous presence of both absorptive and dispersive optical bistabilities in this system with EIT medium inside an optical ring cavity [43]. The output intensity from the optical cavity remains on one of the bistable curves, however, it makes an unexpected switching at the common point E on high intensity branch [4,43]. Of course, a better theoretical model is needed to fully understand such AOM behaviors in this combined absorptive plus dispersion regime, which has always been more difficult to study than the limiting case with either absorptive or dispersive effect alone. It would be very interesting to find a physical explanation of why the cavity intensity switches between the dispersive and absorptive AOB curve at point E. The experimentally observed transitions from AOB to AOM in this system of three-level atoms inside an optical ring cavity are quite different from the previously studied atomic systems for the observation of AOM, since here the transition processes can be controlled and monitored by simply changing one experimental parameter, either the vapor cell temperature or the control beam frequency detuning. Also, this three-level atomic system is much simpler than the multi-Zeeman levels and complicated pumping mechanisms. Optical multistability in conjunction with certain phase coherent techniques [128] can be used to make improved optical transistors, memory elements, and all-optical logic gates. The experimental demonstrations of controlling such OB/OM behaviors can be very useful and significant in realizing such opto-electronic devices in practice.

Chapter 5

Dynamical Instability to Chaos

5.1 Simple Model

A dynamical system can be considered as a stable one if its variables come back to their original values after they are disturbed [129]. The reverse is true for an unstable system. A simple example is the temperature of a refrigerator. It is assumed to be stable because opening its door is a disturbance which will change its temperature but the thermostatic control brings its back to its original set temperature. The stable systems try to maintain themselves in a state which is quite balanced in changing environment of system parameters. The feedback mechanisms in such systems are very critical to determine their stability. For example, the role of negative feedbacks is usually considered to maintain the stability of the system. Such mechanisms control the variables in a manner so that they return toward their equilibrium values like a particle in a parabolic shape well. In other words there are some kind of restoring forces in the system provided by the negative feedbacks. Thus, negative feedback mechanisms provide equilibrium conditions in mechanical, optical, electronic and ecological systems. The presence of negative feedback in a system is a necessary condition for its stability to be maintained but it may not always give a guaranteed stability. The general criterion to ensure the stability of a system with negative feedback is that it should be fast and smooth on the variables of the system which are changing [129].

When a dynamical system drifts away from its original settings once it is disturbed and cannot return back to the initial state, it is called an unstable system. Such unstable behavior of the system has one of its roots in positive feedback because a positive feedback further enhances the changes

127

occurred on the initial state. Hence the stability of a dynamical system is critically dependant on the nature of feedback associated with the system. The stability and instability of systems can be illustrated (Fig.5.1) by a simple consideration where a ball rests on different physical structures or potentials. Figure 5.1(a) depicts the situation of a stable equilibrium with a ball in potential well. In this situation the ball will always return towards its equilibrium position after any kind of disturbance. For an infinitely deep well structure, the system is said to be in a globally stable condition. The situation of unstable equilibrium is shown in Fig.5.1(b). The ball leaves its equilibrium position once it is disturbed and will not return. In Fig.5.1(c), a neutral-stable equilibrium is shown. The ball will stay in its position after external disturbance. Figure 5.1(d) is the situation for metastable equilibrium. In this situation the ball comes back to its equilibrium position provided it is not disturbed too much. This is not a globally stable equilibrium because when the ball crosses the unstable equilibrium position at the peak of the barrier due to disturbance, it does not come back to the stable position again. Finally Fig.5.1(e) depicts another kind of metastable equilibria having multiple stable states. The ball moves back to one or the other equilibrium position depending on its initial position, i.e., towards which side of the unstable equilibrium it is situated initially and a large disturbance can make the ball jump back and forth between the two metastable states. In this situation the system is said to be globally stable [129]. The last situation (e.g., the double well system in Fig.5.1(e)) is

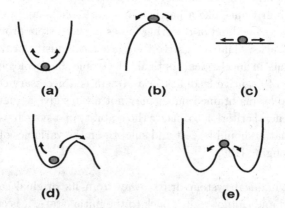

Fig. 5.1 Various situations for equilibria: (a) Stable equilibrium; (b) unstable equilibrium; (c) neutral-stable equilibrium; (d) metastable equilibrium; and (e) metastable equilibria with several stable states (adopted from [129]).

particularly important to our studies of optical bistability since the bistable states in the output intensity are double well system, as will be discussed in later chapters.

5.2 Dynamics of a Coupled Cavity and Two-level Atomic System

The phenomena of AOB and AOM discussed in previous chapters display nonequilibrium phase transitions. The phenomenon of spontaneous temporal oscillations or self-pulsing in the output intensity of AOB, due to instabilities in the system, is considered to be one of the important findings in the subject area of AOB [1,2,130]. Historically, the self-pulsing effect was first discovered in the studies of homogeneously-broadened laser systems. The phenomenon of self-pulsing could be useful to shed light on the general theory of synergetics, which is related to self organizations of structures and patterns in natural systems [131]. More specifically these natural systems are open thermodynamic ones far away from equilibrium. The phase transitions in thermodynamics are related to an order parameter that was generalized by Haken to the 'slaving' principle (SP) [132]. The self-pulsing in AOB has a good agreement with SP, which also explains how enhanced instabilities and bifurcations lead to breaking of the regularity [130]. The simplest way to understand self-pulsing in a coupled cavity with two-level atomic system is to consider the Rabi frequency of the optical field greater than the mode spacing of the cavity, the gain at next mode becomes unstable and it oscillates with a time period that equals to the cavity round-trip time T_c [2]. Another important effect of self-pulsing phenomenon could be in the studies of chaotic dynamics [130] of several physical, chemical, biological, environmental and geological processes. The Ikeda instability [133] occurs when medium response time is much shorter than the cavity round-trip time T_c, which is due to the presence of many simultaneously unstable field modes in a dispersive system. So, the generation of self-pulsing is roughly manifested as a sequence of square waves, culminating into period doubling bifurcations that eventually leads to chaos [2,130].

There are two kinds of instabilities mentioned in the literature in the context of AOB [134]. They are called single-mode and multimode instabilities, respectively. When the resonant mode of the cavity, meaning the longitudinal mode of the cavity nearest to the frequency of the driving input field, becomes unstable, then it gives rise to single-mode instability. However, in the case of multimode instabilities, the side modes of the resonant cavity mode become unstable [134]. The single-mode dynamical instability in the upper branch of a hysteresis curve in two-level AOB system was first studied within the framework of a plane-wave model [130] and later extended to the Gaussian field model [135], which provided the experimental feasibility of the theoretical model. This is because the transverse variation of electric field is quite essential in observing instability which is available through Gaussian field. Dynamical instability was experimentally demonstrated in the upper branch of the AOB curve in a system with two-level sodium atoms inside an optical cavity [15,16]. Such a dynamical effect was shown to connect with the Ikeda instability [133] mechanism, which appears in analyzing the stability of the steady states in OB curves, under the condition that the medium response time is much faster than the cavity round-trip time T_c. Under certain parametric conditions and within the above discussed limit, a sequence of period-doubling bifurcations leading to a region of apparently aperiodic dynamical oscillation, was then observed [15,16]. In the two-level AOB system, multimode instabilities, as well as oscillatory instabilities culminating into optical turbulence, were also reported [136]. In the mixed absorptive-dispersive AOB regime, the off-resonant-mode instability was investigated and it was shown that a part of the lower transmission branch could also be unstable in addition to the upper-branch instability found in the system with pure absorptive bistability [137]. Instability was also observed due to the onset of the cavity side modes in a homogeneously-broadened two-level AOB system [134]. Some other types of self-oscillations and instabilities using different two-level atomic systems inside an optical cavity were also reported in eighties [138].

5.2.1 *Theoretical model of dynamical instability in a two-level system*

The phenomenon of AOB has been studied using models containing two- or three-level medium in either a ring cavity or a FP cavity, as discussed in Chapters 2 and 3. The electromagnetic modes in the optical cavity have spatial variations in general. If the spatial variation of the electromagnetic field modes is not taken into account in the calculations, the problem of AOB can normally be treated in the mean field theory [1,86]. In the mean field theory, several steady-state and some transient analysis have been reported in the literature [1,139]. The plane-wave model is too simple to address issues of low-threshold instabilities in homogeneously-broadened media. However, under appropriate conditions of parameters, one can find this model not only showing periodic behavior, but also giving rise to self-pulsing and/or chaotic phenomena through a sequence of period-doubling bifurcations or via other routes [130]. Note that such a chaotic behavior arises under the conditions of a one-mode theory which is very much different from other similar works, e.g., Ikeda's work [133].

In the following, the theoretical treatment of Ref.[130] will be adopted to illustrate the dynamical instabilities and chaos in the two-level AOB system. The model consists of two-level atoms interacting with a single-mode field inside a ring cavity. We consider only the homogeneous-broadening mechanism in the intracavity medium for the sake of simplicity. The other restrictions imposed in the model are $\alpha L << 1$, $T << 1$, $\delta_0 << 1$, with $C = \alpha L/2T$, $\Theta = \delta_0/T$ (where both C and Θ take some arbitrary values). Here, αL is the unsaturated absorption in the medium of length L, parameter $\delta_0 = \omega_{cav} - \omega_P$ describes the cavity frequency detuning (defined as the difference of external driving field frequency and the nearest cavity resonance frequency), and T is the intensity transmittivity coefficient of mirrors of the ring cavity where from light is entering and coming out. In order to enumerate the dynamical evolution of atom and cavity field, one has to employ the Maxwell-Bloch equations along with the appropriate boundary conditions and physical conditions mentioned above [130]. It was shown in Ref.[140] that these equations can be mapped onto an infinite hierarchy of coupled equations for the complex field amplitudes, polarization and population variables. It is possible to simplify these equations further by assuming all atomic and field modes are spatially uniform and no contribution from the nonresonant modes during the evolution. The simplified Maxwell-Bloch equations reads as [130]

$$\dot{x} = -\kappa[i\Theta x + (x - y) + 2CP],$$
$$\dot{P} = \gamma_\perp[xD - (1 + i\Delta)P],$$
$$\dot{D} = -\gamma_\parallel[\frac{1}{2}(xP^* + x^*P) + D - 1], \qquad (5.1)$$

where κ is the cavity field decay rate, $\gamma_\perp$ represents, decay of the atomic polarization and $\gamma_\parallel$ is the rate of decay of the population difference. The parameter Δ represents the atomic detuning defined as $(\omega_{12} - \omega_P)/\gamma_\perp$. The variables x, y, P and D are parameters directly related to the complex transmitted field, the input field, the atomic polarization and the difference in population of two levels as defined in the following (in terms of the original Eq.(2.1) of the Chapter 2): $D = \frac{1}{2}(\rho_{22} - \rho_{11})$, P is proportional to ρ_{12}, $\gamma_\| = \Gamma_{21}$ and $\gamma_\perp = \gamma_{21}$. The transmitted field x, input field y, saturation intensity I_s, and α are defined in section 2.1. For the sake of simplicity, the polarization relaxation time is considered to be the smallest among all the characteristic times related to the system evolution, e.g., $\gamma_\perp >> \gamma_\|, \kappa, \kappa\Theta, \kappa C$. This makes it possible to eliminate the polarization adiabatically, i.e., it is straightforward to solve polarization equation in the steady-state condition and substitute the steady-state value of P in remaining equations. In this way the variable P is eliminated adiabatically. By doing so the following coupled equations are obtained [130]

$$\dot{x} = -\kappa[i\Theta x + (x - y) + 2C\frac{xD}{1 + i\Delta}],$$

$$\dot{D} = -\gamma_\|[\frac{D|x|^2}{1 + \Delta^2} + D - 1]. \tag{5.2}$$

The parameter x is a complex quantity, so Eq.(5.2) contains three first order differential equations. The stability analysis of these equations can be performed using a standard procedure, i.e., by linearizing them around a steady-state solution. By defining $x(t) = x_{ss} + \delta x(t)$ and $D(t) = D_{ss} + \delta D(t)$, and using them in Eq.(5.2), it is easy to observe that quantities like δx, δx^*, and δD, which represent deviations from the steady-state will evolve in time as linear superpositions of exponential quantities [130]

$$\delta x(t) = \sum_{i=1}^{3} g_i \exp(\Lambda_i t), \tag{5.3}$$

where Λ_i $(i = 1 - 3)$ are solutions of the cubic equation $\Lambda^3 + b_1\Lambda^2 + b_2\Lambda + b_3 = 0$. The coefficients of cubic equation: b_1, b_2, and b_3 are functions of the parametric conditions set in the system. These coefficients b_i $(i = 1-3)$ are real and further assuming that Re $\Lambda_i < 0$ $(i = 1 - 3)$, then the stability analysis can be obtained using the Hurwitz criterion [130,131]. The Hurwitz criterion demands that coefficients b_i $(i = 1 - 3)$ to satisfy stability conditions. It is easy to show that the Hurwitz stability conditions $b_1, b_2 > 0$ are clearly satisfied for all parametric values. If $dy/dx > 0$, then $b_3 > 0$ is also satisfied. Note that the condition $b_1 b_2 > b_3$ is very important because it gives some basic clues for the occurrence of the instability in the upper transmission branch for a given choice of the parameters C, Δ, Θ and $\kappa/\gamma_\|$. However, for large values of C, a wide range of instability would show up in the upper branch [130].

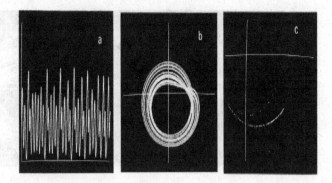

Fig. 5.2 Typical chaotic self-pulsing behavior for $y = 950$, $\Delta = 374$, $\Theta = 340$, $C = 70,000$. (b) Phase-space representation in the (Re x, Im x) plane. (c) Poincaré map of the solution obtained by intersecting the three-dimensional trajectory with a plane $D = 0.845$ (adopted from [130] with permission).

Using the above formulated plane-wave theory of AOB, several interesting single-mode pulsation phenomena have been suggested in the theoretical investigations [1,2,16,130]. Note that these phenomena are different from the dynamical effects arising in the cavity due to multimode fields. When atoms and cavity were in exact resonance condition: $\Delta = 0$, $\Theta = 0$, no positive-slope instability [1,2,16,130] could be investigated. Instability in the system occurred due to the non-zero cavity detuning. The instability could occur with and without AOB. The development of instability was not so difficult when the cavity detuning and the atomic detuning had opposite sign.

The output field intensity, when plotted as a function of time displays a non-periodic behavior, as shown in Figs.5.2(a). This is also evident in Fig.5.2(b) in which phase-space plot of Re(x) vs Im(x) is given. These plots are for the parametric condition: $\Delta = 374$, $\Theta = 340$, $C = 70,000$. To get insight into the periodic or chaotic type of solutions, Poincaré maps of the trajectories in the three-dimensional phase-space were analyzed [130]. These Poincaré maps were acquired by getting the real and imaginary parts of the field amplitude x at the place where the phase-space trajectory crosses a prespecified plane characterized by the condition $D = $ constant, once in every loop. The Poincaré maps are generated from aperiodic solutions of regular-looking lines (Fig.5.2(c)) and the individual points spread on the map in a random manner [130].

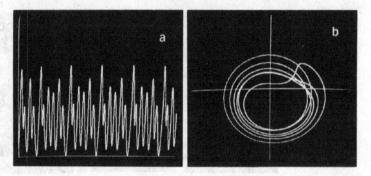

Fig. 5.3   An example of a periodic solution (6P) in a window of periodicity contained within the chaotic domain. Several such windows have been observed. (a) Time-dependent trajectory; (b) trajectory in the (Re x-Im x) phase-space (adopted from [130] with permission).

There are certain ranges (or windows) of parametric value of y (within the chaotic domain) for which output intensity again shows a periodic behavior. This is shown in Figs.5.3(a,b) where a periodic solution (6P or period 6) is shown. Once we come out from chaotic region, there is inverted sequence of bifurcations giving rise to simple oscillatory self-pulsing and culminating to precipitation. However, there is some difficulty to predict that the sequence has an accumulation point because of the uncertainty in numerical solution. The aperiodic solutions show the expected behavior of chaotic structures (at least qualitatively) in the system but the important issue is that these solutions depend critically on the accumulated computational error [130]. The solutions were a sensitive functions of different computer codes and step sizes. At some stage during the integration, different codes gave different results which were quite apart from each other. However, the detailed structure of the solution was well reproduced for decent lengths of time. One does not encounter such problem when there is a periodic behavior in oscillations. It is very interesting to note that the Poincaré maps are independent of the details of numerical methodologies used [130]. The model discussed above is quite general in the sense that it can also deal with a Kerr medium [141] if one defines $D = 1 - e(t)$, and impose the conditions $e(t) << 1$ and $2C << \Delta^2$. The physical meaning of these conditions are as follows: $e << 1$ clearly states that there is a negligible amount of absorption of the incident energy in the medium. The other condition $2C << \Delta^2$ guarantees that when the intracavity medium is getting bleached ($\theta = 2C\Delta/(1 + \Delta^2 + |x|^2)$), there is no reflection of light [130]. In other words the expression $|x|^2 = y^2$ holds good. Under such approximations, Eq.(5.2) provides equations similar to ones discussed in Ref.[141]. However, the model discussed here is not always valid for all the parametric conditions discussed in [141].

5.2.2 *Experimental demonstrations of single-mode instability in the two-level system*

Single-mode instability was experimentally demonstrated in a setup similar to the one used for demonstrating AOB in two-level system but with certain improvements and modifications [15,16]. This kind of single-mode instability could be considered as a passive counterpart of the Lorentz-Haken model which describes the active instability. The experimental arrangement for observing AOB in the two-level sodium atomic system is described in Chapter 2. The appearance of instability occurs only when both the optical cavity and atoms are off-resonance with the intracavity electromagnetic field frequency. Hence the precise knowledge of the atomic and the cavity detunings were quite important in these experiments [15,16]. During the experiment the dye laser beams and atomic beams were aligned perpendicular to each other to make sure the best possible elimination of the first-order Doppler-broadening. The retro reflected light beam from a corner cube prism was employed to achieve this goal [15,16]. The weak-signal absorption calibration was performed and it was correlated to the optical-pumping fluorescence. Once alignment and calibration were completed, the atomic number density was changed by varying the temperature of the atomic oven. The light going inside the cavity was modulated at a frequency of 50 Hz approximately. When compared with the dynamical rates of the system, the modulation frequency was several orders of magnitude lower. Monitoring of powers were continuously done for both ingoing as well as outgoing light from the cavity.

The definitions used for the cavity detuning and atomic detuning are as follows: $\Delta = \frac{\omega_{21} - \omega_P}{\gamma_\perp}$, $\Theta = \frac{\omega_{cav} - \omega_P}{\kappa}$, in which ω_P is the incident or probe (cavity) field frequency, ω_{cav} is the cavity mode frequency, and ω_{21} is the frequency of atomic transition of interest. The cavity field decay rate and polarization decay rate are κ and $\gamma_\perp$, respectively. The setting of atomic and cavity detunings were done in the following manner to see the development of instability. The ratio of lower and upper switching threshold powers in the resonant absorptive AOB case was monitored and when it became about two, both the atomic and cavity detunings were changed from zero but keeping $\Delta\Theta < 0$. Once the instability was developed, there was an extraordinary change in the input-output characteristics of the cavity. This could be understood by the fact that the average value of the pulsating state was moved above the corresponding time-independent steady-state. A good agreement between the theoretical prediction and the experimental results were found. For a given value of C but changing the cavity and laser detunings, the instability boundary in the (Θ, Δ) plane was investigated. By doing so the absolute comparison between experimental results and theoretical calculations were possible. Figure 5.4 shows the instability boundaries explored in the atomic detuning vs cavity detuning space (Θ, Δ) of the system. The figure also provides the absolute quantitative comparison of the theoretical predictions with experiments. These results are for a optical ring cavity in which the cooperativity parameter C is directly related to the atomic number density [15,16]. During the experiment, measurements were taken by keeping atomic detuning fixed but keep changing cavity detuning. The line broadening mechanism, prominent in experiments was the transit broadening mechanism, which was considered to be a homogeneous-broadening mechanism. The presence of this mechanism gave rise changes in the polarization decay as well as the saturation intensity. Such a assumption does not have any well defined reason, but it provided a good matching of the experimental results with theoretical

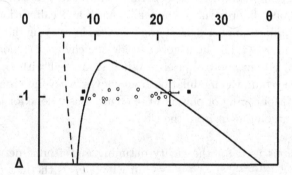

Fig. 5.4   Domain of instability in the detuning space for $C = 60 \pm 5$, $\kappa' = 0.5$, and $\gamma = 1.6$. Open circles indicate instability, closed squares stability. Below the continuous line is the unstable region based on the Gaussian theory, while to the right of the dashed line is the plane-wave prediction of the unstable region. The ring cavity is used in this measurement. Systematic uncertainties in the detuning are ± 10 % (reprinted from [16] with permission).

predictions, when switching points (i.e., threshold points) along the lower and upper branches were calculated [16] in the resonant absorptive AOB situation. The approximation came out with a very well quantitative agreements with experiments when both atomic as well as cavity detunings were small [15,16]. Since atomic beams could not be collimated perfectly, there was a good possibility of inhomogeneous broadening (of the order of 1 MHz) in the system but that was neglected. The plane wave theory of AOB was used in the theoretical calculation and was not in good agreement with the experimental data in general, as shown in Fig.5.4. This is because the instability domain is quite wide and there was an uncertainty in getting the uniform field while carrying out experiments. The work carried out here for the selected parametric conditions was in good agreement with other works on instabilities [15,16] under the nonuniform field structure. Some other experiments were carried out in the following ranges of parameters: $0 < |\Delta| < 6$, $0 < |\Theta| < 40$, $C < 350$ [15,16] to explore the self-pulsing phenomenon. The resonator decay rate was smaller than the polarization decay rate (the ratio of two decay rates was about $1/2$), but no higher-order bifurcation of the self-pulsing states could be observed in the upper branch of the AOB hysteresis curve. Due to the presence of some kind of gain-assisted generations of counter-propagating waves (the lasing action), the cavity field circulated in either directions in the ring cavity. The counter-propagating waves had higher-order transverse mode structures and not the usual TEM_{00} mode. Sometimes it was possible to make correspondence of

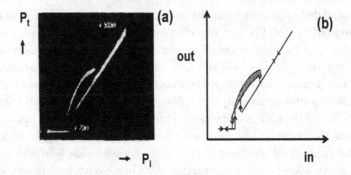

Fig. 5.5 (a) Input-output characteristics (P_i vs P_t) for $C = 250 \pm 50$, $\Theta = 14 \pm 3$, $\Delta = -3 \pm 1.0$, $\kappa' = 0.4$, and $\gamma = 1.6$ in the ring d cavity, when hard-mode instability is present, giving rise to a new bistable region associated with the oscillatory state. (b) Explanatory drawing of the instability as an arch separated from upper branch, the arrows indicating the the direction of switching as the input intensity is scanned at a rate of 50 Hz (adopted from [16] with permission).

the counter-propagating beam with one of the multiple branches of AOB hysteresis curve. In Fig.5.5, a typical AOB hysteresis curve is shown for the unstable region. Note that the effect shown in Fig.5.5 is due to the hard-mode excitation, i.e., the finite magnitude perturbation is responsible for the system to become unstable. The instability can be interpreted in terms of a new hysteresis cycle associated with oscillatory states. Also, the role of self-focusing and the mode reshaping was ruled out in the experimental observation of such single-mode instability in AOB hysteresis cycle [16].

The conclusions of such experiments were that the formulation based on Gaussian model of the field were valid even in the regimes of spontaneous oscillations, i.e., the output field remains Gaussian in such conditions. Also, the instability boundaries in the parametric space and the oscillation frequency are both matching well with the predictions of the Gaussian model. The model based on plane-wave assumption was not good enough to explain the experimental findings [16].

5.3   Dynamical Instabilities in Three-level Atomic Optical Bistability

Instability in AOB was also observed in the transmission field of an optical cavity containing cold atomic clouds of cesium, where dynamics of the system was governed by the degenerate Zeeman sublevels [17]. This instability was slightly different from the previously observed ones. In the experiment reported in Ref.[17], just one circularly-polarized laser beam was employed going into the optical cavity and interacting with all the Zeeman sublevels of the $6S_{1/2}F = 4$ and $6P_{3/2}F' = 5$ states in cesium atoms. The origin of dynamic instability was due to the competitive processes of (a) the optical pumping to state $6S_{1/2}F = 4$, $m_F = 4$ from all other Zeeman sublevels and (b) nonlinearity caused by the saturation of the optical transition from the state $6S_{1/2}F = 4$, $m_F = 4$ to the state $6P_{3/2}F' = 5$, $m_{F'} = 5$. The oscillations obtained in the cavity transmission field was modeled as a quasi-two-level system interacting with one cavity field mode, which could only explain the experimental results qualitatively [17]. In the experiment both optical pumping and saturation effects within the optical cavity were created by the same input laser beam, hence there was no separate control over the two competing dynamical processes. Other limitations for fully understanding instabilities in that experiment were due to the presence of degenerate Zeeman sublevels and the trapping and repumping beams, required to prepare cold clouds of cesium atoms.

The major issue in studying optical dynamical instabilities involving complicated energy structures in various atomic systems has been the lack of experimental controls to investigate such dynamical effects in a systematic manner and correlate them with quantitative theoretical calculations. This issue of controllability could be removed by using a simple three-level EIT system inside an optical ring cavity. Using such EIT systems one can easily control the various linear and nonlinear interaction processes in the intracavity medium with the frequencies and intensities of the probe and coupling (control) laser beams over a wide range of values. To observe such dynamical oscillation (self-pulsing) phenomenon, relatively very low laser intensities are needed and the oscillatory behavior depend very sensitively to the frequency detunings and intensities of the two independently applied probe and coupling laser beams. Since the linear and nonlinear optical properties of simple three-level EIT medium are well understood, a better theoretical model can be developed to provide quantitative comparisons with the experimentally measured results, which is necessary to fully understand such interesting instability behaviors in interacting atom-cavity systems.

5.3.1 *Experimental demonstration of instability in a three-level EIT medium*

The experimental setup to control instability in a Λ-type three-level atomic system is basically the same as the one described earlier for observing and controlling AOB (in Chapter 3) and AOM (in Chapter 4), as shown in Fig.3.14 in an optical cavity containing rubidium atomic vapor. An atomic cell containing Λ-type three-level atoms is placed inside a three-mirror optical ring cavity, and an EOM is used before the cavity input to adjust the intensity of the probe (cavity) field. Although AOB exists in such experimental setup, it was not specifically studied during the following described experiments. The atomic transitions used in this instability experiment are shown in Fig.1.9(b), where a typical Λ-type configuration of ^{87}Rb atomic levels is displayed. Before starting the main experiment, the probe laser frequency and the coupling laser frequency were both locked to their respective atomic transitions with the help of two separate FP cavities dedicated for this purpose. Small frequency detunings for both the probe and coupling lasers could be achieved by changing the applied voltages on the PZTs of the dedicated FP cavities used to lock the lasers [46].

The scanning of the optical ring cavity across its resonance was achieved through a ramp voltage applied to the PZT on which mirror $M3$ was mounted. An APD was employed to measure the light transmitted from the cavity. The cavity transmission profile of the ring cavity (measured by scanning the cavity length) was quite symmetric in shape in the absence of the coupling (control) laser beam. The cavity transmission profile underwent a significant change in the presence of the coupling beam and became asymmetric and/or displayed a ringing (or oscillatory) phenomenon due to the generation of dynamical instability in the system. These dynamical oscillations or the ringing phenomenon in the cavity transmission profile were generated due to the competition between the saturation in the probe transition and the optical pumping effect in the coupling transition, which are enhanced in the EIT medium due to induced atomic coherence. These optical processes are determined by the absorption and nonlinear properties in the interaction between the optical fields and the respective atomic dipoles in the two transitions. Such qualitative results were discussed in an initial experimental work [142]. According to that work [142] the refractive index of the medium inside the optical cavity undergoes rapid changes due to the control field induced optical pumping, which increases the refractive index and the saturation caused by the intracavity field decreases it. The end result of such changes in refractive index gives fluctuations in the cavity length so that the cavity field alternates between two different states of the AOB quite rapidly, which is exhibited as self-pulsing phenomenon [142]. Therefore such dynamic instability phenomenon is very sensitive to the experimental conditions, i.e., intensities and frequency detunings of the coupling as well as probe laser beams. The measured dynamic instability as functions of these experimental conditions are discussed next.

The experimental observations of optical dynamic oscillation in the cavity transmission profile are presented in the left column of Fig.5.6 for three different probe frequency detunings. The other parameters selected in the experiment were as follows: coupling laser power P_c =11 mW, probe laser power P_p^{in} = 2.8 mW, and $\Delta_c = 0$. The probe frequency detuning was varied and it was kept $\Delta_p = 25$ MHz in plot (a); $\Delta_p = 35$ MHz in plot (b); and $\Delta_p = 45$ MHz in plot (c). The measurement of the input power entering the cavity was done just at the entrance mirror of the cavity [5,46].

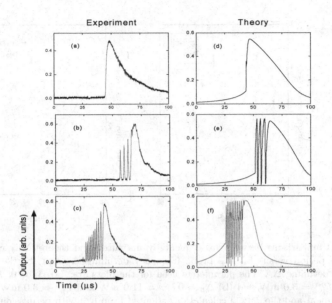

Fig. 5.6 Experimentally observed and theoretically calculated cavity transmission profiles from an optical ring cavity containing three-level Λ-type atoms for three different values of probe frequency detuning. The left column is the experimental observations and the right column is the corresponding theoretical calculations. The parameters used in the experiment and theoretical calculation are: $P_c = 11$ mW, $P_p^{in} = 2.8$ mW, $\Delta_c = 0$, and (a,d) $\Delta_p = 25$ MHz; (b,e) $\Delta_p = 35$ MHz; (c,f) $\Delta_p = 45$ MHz (reprinted from [46] with permission).

Another interesting study is related to the experimental investigation of the variation of measured period of oscillation as a function of probe beam frequency detuning Δ_p that has been plotted in Fig.5.7(a) (the dash curve). For the plot, parametric conditions used in experiments are $\Delta_c = 0$, $P_c = 11.0$ mW, $P_p^{in} = 3.0$ mW, and the atomic cell temperature $T = 70.3$ °C. Plot of experimentally measured oscillation period as a function of coupling frequency detuning (Δ_c) is given in Fig.5.7(b) (the dash curve). The parameters used in this study are $\Delta_p = 0$, $P_c = 11.0$ mW, and $P_p^{in} = 3.0$ mW. The next noteworthy study on such dynamic oscillation is to find the effect of the probe and coupling (control) laser powers on the time period of oscillations. The measured oscillation time period in the experiments as a function of the control laser beam power is displayed in Fig.5.8(a) (the dash curve). Experimental parameters used in this plot are: $\Delta_c = 0$, $\Delta_p = 35$ MHz, $P_p^{in} = 3.0$ mW, and $T = 70$ °C. On the other hand, in Fig.5.8(b), the experimental curve for the oscillation period as a

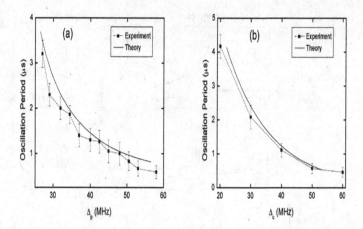

Fig. 5.7 Comparisons between experimentally measured and theoretically calculated oscillation period versus (a) the probe frequency detuning Δ_p and (b) the coupling frequency detuning Δ_c. The parameters used for the plots are: (a) $\Delta_c = 0$, $P_c = 11.0$ mW and $P_p^{in} = 3.0$ mW and (b) $\Delta_p = 0$, $P_c = 11.0$ mW and $P_p^{in} = 3.0$ mW for both the experiment and theoretical calculation (adopted from [46] with permission).

function of cavity input power is displayed (the dash curve). The values of parameters used in this situation are: $\Delta_c = 0$, $\Delta_p = 35$ MHz, $P_c = 11.0$ mW, and T $= 70$ °C. All these experiments were repeated several times under the similar experimental parametric conditions [5,46].

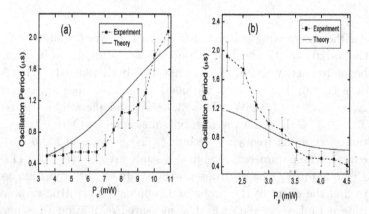

Fig. 5.8 Comparisons between experimentally measured and theoretically calculated oscillation periods versus (a) the coupling power and (b) the cavity (probe) power. The parameters used are: (a) $\Delta_c = 0$, $\Delta_p = 35$ MHz and $P_p^{in} = 3.0$ mW and (b) $\Delta_c = 0$, $\Delta_p = 35$ MHz and $P_c = 11.0$ mW for both experiments and theoretical calculations (adopted from [46] with permission).

5.3.2   *Theoretical calculation*

For the purpose of theoretical calculation, the model consists of a system of three-level Λ-type atoms (number density N) inside an optical ring cavity interacting with the probe (cavity) and coupling laser fields, as shown in Fig.1.9(b) of Chapter 1. The theoretical model is based on the density-matrix equations for three-level atoms (Eq.(1.58)) and the modified Maxwell's equation for the cavity field, e.g., the input-output field relationship equation (Eq.(3.13)), along with Eq.(3.16)). The exact analytic solution of the density-matrix equations is beyond the reach. Steady-state solution for the density-matrix equations can be obtained but that is of no use in the present context since the instability behavior is time dependent. So, in order to obtain some meaningful results, the exact numerical simulation has to be carried out using the atomic equations (Eq.(1.58), Chapter 1) along with cavity field equations, both of which are written in slightly different forms as given in the following [46,47]

$$\dot{z}_p = -\frac{\gamma_{23} + 2\gamma_{21}}{3}(1 + z_p + z_c) - \gamma_{31}(z_p - z_c)$$
$$\quad - 2x_p(\alpha_1 y_{21} + \alpha_2 x_{21}) - x_c\alpha_c y_{23},$$

$$\dot{z}_c = -\frac{2\gamma_{23} + \gamma_{21}}{3}(1 + z_p + z_c) + \gamma_{31}(z_p - z_c)$$
$$\quad - x_p(\alpha_1 y_{21} + \alpha_2 x_{21}) - 2x_c\alpha_c y_{23},$$

$$\dot{x}_{21} = -\gamma x_{21} + \Delta_p y_{21} + 2x_p\alpha_2 z_p + x_c\alpha_c y_{31},$$

$$\dot{y}_{21} = -\gamma y_{21} - \Delta_p x_{21} + 2x_p\alpha_1 z_p - x_c\alpha_c x_{31},$$

$$\dot{x}_{23} = -\gamma x_{23} + \Delta_c y_{23} - x_p(\alpha_1 y_{31} + \alpha_2 x_{31}),$$

$$\dot{y}_{23} = -\gamma y_{23} - \Delta_c x_{23} + 2x_c\alpha_c z_c - x_p(\alpha_1 x_{31} - \alpha_2 y_{31}),$$

$$\dot{x}_{31} = -\gamma_{31}x_{31} + (\Delta_p - \Delta_c)y_{31} + x_c\alpha_c y_{21} + x_p(\alpha_1 y_{23} + \alpha_2 x_{23}),$$

$$\dot{y}_{31} = -\gamma_{31}y_{31} - (\Delta_p - \Delta_c)x_{31} - x_c\alpha_c x_{21} + x_p(\alpha_1 x_{23} - \alpha_2 y_{23}),$$

$$\tau\dot{\alpha}_1 = t_1\alpha_p^{in} - \gamma_{cav}\alpha_1 - \Delta_\theta\alpha_2 - [1 + \frac{2}{3}(1 + z_p + z_c)]e\alpha_2 - f(\alpha_1^2 + \alpha_2^2)\alpha_2,$$

$$\tau\dot{\alpha}_2 = -\gamma_{cav}\alpha_2 + \Delta_\theta\alpha_1 + [1 + \frac{2}{3}(1 + z_p + z_c)]e\alpha_1 + f(\alpha_1^2 + \alpha_2^2)\alpha_1,$$

$$\text{(5.4)}$$

where $z_p = \rho_{22} - \rho_{11}$ is the population inversion between the level $|2\rangle$ and level $|1\rangle$ and $z_c = \rho_{22} - \rho_{33}$ is the population inversion between the level $|2\rangle$ and level $|3\rangle$. The quantities x_{ij} and y_{ij} $(i, j = 1 - 3)$ are the real and imaginary parts of the corresponding density-matrix elements ρ_{ij} $(i \neq j)$, respectively [46,47]. The probe and coupling frequency detunings are defined as $\Delta_p = \omega_p - \omega_{21}$ and $\Delta_c = \omega_c - \omega_{23}$, respectively, as before. Δ_θ is the cavity frequency detuning. The radiative decay rates of the system are as follows: γ_{21} is the radiative decay rate from level $|2\rangle$ to level $|1\rangle$ and γ_{23} is the radiative decay rate from level $|2\rangle$ to level $|3\rangle$. The nonradiative decay rate from level $|3\rangle$ to $|1\rangle$ is γ_{31}. The quantity γ is related to decay of coherence and given by $\gamma \equiv (\gamma_{21} + \gamma_{23} + \gamma_{31})/2$. The Rabi frequencies corresponding to the probe and coupling beams are defined as $\Omega_p = -\frac{\mu_{21}}{\hbar}\sqrt{\frac{\hbar\omega_p}{2n\epsilon_0 c S_p}}\,\alpha_p \equiv -2x_p\alpha_p$, and $\Omega_c = -\frac{\mu_{23}}{\hbar}\sqrt{\frac{\hbar\omega_c}{2n\epsilon_0 c S_c}}\,\alpha_c \equiv -2x_c\alpha_c$, respectively, where μ_{21} (μ_{32}) is the transition dipole matrix element connecting levels $|2\rangle$ and $|1\rangle$ (levels $|2\rangle$ and $|3\rangle$). The quantity n is the refractive index of the medium. The parameters S_p and S_c are the areas of the probe and coupling (control) beams, respectively. The complex probe field strength is given by $\alpha_p = \alpha_1 + i\alpha_2$, and $|\alpha_p|^2$ is related to the average number of photons flowing (expressed as number of photons per second [46,47]). The coupling field strength is defined by α_c, which does not get on resonance with the cavity. The quantity α_c can be considered as a real number without losing any generality. The two equations given at the end of Eq.(5.4) describe the dynamical evolution of the intracavity field (real and imaginary parts) and they are coupled differential equations. It should be noted that only the probe field α_p circulates inside the cavity as the intracavity field, which makes a round-trip in the optical cavity during a time interval τ. During this time interval there are changes introduced in the intracavity field because of following reasons: (i) A driving field α_p^{in} enters the cavity via mirror M1 (having a transmission coefficient t_1). (ii) The intracavity field decays with a decay constant γ_{cav} due to finite transmission of mirrors and losses incurred in the intracavity medium (for an EIT system, the losses due to intracavity medium are reduced). (iii) By the round-trip (linear and nonlinear) phase shift $\Phi_{cav} = \Phi_0 + (1 + \frac{2}{3}(1 + z_c + z_p))e + f|\alpha_p|^2$, where Φ_0 specifies the phase shift acquired due to free space field propagation in the cavity, and variables e and f are defined in the following. This phase shift is determined by the atomic variable equations in Eq.(5.4) [5,46].

The parameter Δ_θ is related to the cavity detuning in these two field equations. The quantities e and f are dependent on the parameters Δ_c, Δ_p and thus related to the linear and the nonlinear phase shifts introduced by the cavity. The expressions for e and f read as [5,46]

$$e = Ng^2/\Delta_p, \tag{5.5}$$

$$f = -2Ng^4/\Delta_p^3, \tag{5.6}$$

where g is the coupling constant of atoms with the field given by

$$g = \sqrt{\frac{\mu_{21}^2 \omega_p}{2\epsilon_0 \hbar S_p c}}. \tag{5.7}$$

These expressions [46] basically agree with the model used in Ref.[17].

With the theoretical model developed above many interesting features on the dynamic oscillation can be predicted and understood. First, a qualitative study comparing the experimental observations (left column) and the theoretical calculations (right column) of the optical dynamic oscillation in the cavity transmission is presented in Fig.5.6 for three different probe field frequency detunings. The other parameters selected in the experimental measurements and theoretical calculations are mentioned in the figure caption of Fig.5.6. The theoretical results are in good agreement with the experimental results. The time period of dynamic oscillation decreases with increasing probe field frequency detuning. The oscillation finally ceases when the probe field frequency detuning reaches a value $\Delta_p = 60$ MHz, which was observed but not shown in the figure here [5,46].

Next, a comparative study of the measured period of oscillation as a function of the probe beam frequency detuning Δ_p, plotted in Fig.5.7(a) with the calculated theoretical curve is given under the same parametric conditions. From the figure it is clear that reasonable agreement has been realized between the experimental observations and the theoretically simulated results. For this plot, parametric conditions used for both experiments and theoretical simulations are $\Delta_c = 0$, $P_c = 11.0$ mW, and $P_p^{in} = 3.0$ mW. The atomic number density value utilized for the simulation of theoretical curve is $N = 10^{10}/cm^3$, which means an atomic vapor temperature of $T = 70.3$ °C in the experiment. In Fig.5.7(b), plots of experimentally measured and theoretically simulated oscillation period as a function of the coupling frequency detuning (Δ_c) are displayed. The parameters used in this study are $\Delta_p = 0$, $P_c = 11.0$ mW, and $P_p^{in} = 3.0$ mW for both experiment and theoretical calculation [5,46].

Another interesting study on such dynamic oscillation is to find effects of the probe and coupling laser powers on the time period of the self-pulsing oscillations. The measured oscillation time period in the experiments as a function of the coupling laser beam power as well as the theoretically calculated results under the same conditions of experiments are displayed in Fig.5.8(a). For both experiments and theoretical calculation following parameters were used: $\Delta_c = 0$, $\Delta_p = 35$ MHz, $P_p^{in} = 3.0$ mW, and T = 70 °C. On the other hand, in Fig.5.8(b), both experimental and theoretical curves for the oscillation periods as a function of the cavity input power are displayed. The values of parameters used here are: $\Delta_c = 0$, $\Delta_p = 35$ MHz, $P_c = 11.0$ mW, and T = 70 °C. In these experiments, all the parameters were measured directly and no fitting parameters are used for making comparisons between the experimental results and the theoretical calculations [5,46].

As discussed above the simple physical basis of such dynamic instability can be understood from the following consideration. There exist two competing dynamical processes for the three-level atomic system inside the optical cavity, i.e., optical pumping from the state $5S_{1/2}$ $F = 2$ to the state $5S_{1/2}$ $F = 1$ by the coupling (control) field and the nonlinear saturation effect in the transition from the state $5S_{1/2}$ $F = 1$ to the state $5P_{1/2}$ $F' = 2$ due to the intracavity probe field circulating in the cavity [5,46,142]. These two dynamical processes have different time characteristics depending on the relative strengths of the two laser beams and the transitions probabilities of two transitions. When the two dynamical processes have comparable strengths, the system can not decide which way to go and therefore becomes unstable. This resembles the situation when two fighters have same power, so they knock each other down as time goes on (as an 'oscillation'). Thus the system shows oscillatory behavior and gives rise to instability phenomenon. The feedback provided by the optical cavity also plays an essential role to create such instability in the cavity field, as determined by the field equations in Eq.(5.4) [5,46]. This observed instability in three-level atomic medium is unique and may not be found in two-level AOB systems. The time period of such dynamic oscillation critically depends on the relative dominance of these two dynamical processes as discussed above, which are governed mainly by the intensities and frequency detunings of the coupling and probe laser beams, as demonstrated in experiments [46,142]. The interesting point to note is that two frequency detunings (Δ_P and Δ_C) play the same role in changing the oscillation period (e.g., the oscillation period decreases as the frequency detuning increases), as shown in Fig.5.7, while the probe and coupling laser powers play the opposite role. To be explicit, as both coupling and probe field frequency detunings go up, there is increased optical saturation. The higher probe power enhances the nonlinear saturation while higher coupling laser power favors the effect of optical pumping. With the increase in coupling laser frequency detuning, the extent of the optical pumping goes down leading to a relative enhancement in the optical saturation process. On the other hand, the cause of increased optical saturation with increased probe field frequency detuning is essentially due to the usual absorption characteristics of EIT medium, which means there is a lower absorption at exact resonance condition due to fulfilment of EIT conditions, but larger absorption under off-resonance condition or nonzero probe detuning [46]. In Figs.5.7 and 5.8 such behaviors can be clearly observed.

In the theoretical modeling certain important effects, e.g., spatial distribution of the intracavity laser field (Gaussian beam profile), focusing, and the propagation effects of the beams in the vapor cell are neglected for the sake of simplicity [5,46]. Also, there is an uncertainty in determining the intracavity intensity from the cavity input power because of some approximate estimation of the beam size for cavity field. The measured and calculated EIT absorption values within the cavity showed appreciable differences and hence the saturation is affected. Another important issue, not considered in the optical pumping process, is the contributions from Zeeman sublevels. The residual second-order Doppler effect is also not accounted for in these calculations. Such simplifications in the calculation give rise significant discrepancies in the two curves (experiment vs theory) of oscillation period versus coupling and probe powers in Fig.5.8. One more limitation in these experiments was the relatively low power of the coupling laser (12 mW as a maximum limit) available at the cavity entrance. The theoretical model has predicted that the period of oscillation will increase initially, up to a maximum value with an increase in the coupling power, but it will decrease towards zero with very large coupling power [5,46]. Such interesting effect in the theoretical prediction could not be tested due to the limited coupling power available in the experiment.

The observation of dynamic instability in the cavity output field shows that this nonlinear system with three-level atoms inside an optical cavity becomes unstable. It has been well established that system in such self-pulsing oscillation regime can be further driven into regions of period-doubling, period-quadrupling etc, and eventually into chaos [47]. The route from period-doubling to chaos can be achieved by adjusting certain system parameter(s). In the following, it will be shown that this composite system with three-level EIT atoms inside the cavity can be driven to chaotic regime by simply tuning the coupling beam frequency detuning Δ_C from a finite value (tens of MHz for observing instabilities) towards zero, near which the linear and nonlinear dispersion slope become very large [25,33].

5.4   Chaotic Dynamics in the Three-level Atomic System

Chaos is one of the most intriguing natural phenomena [143]. In the language of mathematics, the chaotic phenomenon appears in the nonlinear dynamical systems, whose time evolutions are sensitively dependent on the initial conditions. The trajectories of a chaotic system in the appropriate phase space diverge away from each other in exponentially fast manner with a slightly different starting conditions. The phenomenon of chaos suggests disorder and randomness but it is not so because system exhibiting chaos have some kind of order and they are deterministic. Thus the understanding of chaotic dynamics is not only important from the point of view of fundamental interests but also because of its many applications in physical, biological, and chemical sciences. The chaotic phenomena also appear in economics, philosophy, communications (to provide encrypted message [144]), medical sciences (in maintaining normal cardiac function [145]), and weather forecasting. The existences of dynamic instability and chaos were both theoretically predicted in two-level AOB systems [130] as discussed in previous sections. Although the dynamic instability was experimentally observed in several two-level systems consisting of either atoms or molecules [15,146], the predicted chaotic dynamics and route to chaos via period-doubling have not been observed in typical two-level AOB systems, except in a degenerate atomic system with strong radiation trapping [147]. In most cases, the predicted chaos occurs at parametric regions which are not easily accessible in real experiments.

It is clear that certain specific conditions are needed to observe chaotic phenomenon in AOB systems. If the longitudinal and radial variations of electric field are present in the cavity [134], then it is quite likely that chaotic dynamics will not show up. Chaotic dynamics have been experimentally investigated and studied in several laser systems [146,148,149] and in a hybrid optical bistable system [150]. Also, controllability of chaos has been demonstrated by utilizing various feedback mechanisms [151].

By employing the system with three-level atomic EIT medium inside an optical ring cavity, it has been shown that chaos can be reached via period-doubling route in certain parametric regime [47]. By manipulating one of the experimental parameters, such as frequency detuning of the coupling laser beam interacting with the EIT medium, the transmitted cavity field can be driven from self-pulsing with a well-defined oscillation period to period-doubling, period-quadrupling, and finally to chaotic situation [152]. The enhanced linear and nonlinear dispersions around EIT resonance in such a three-level atomic system are the prime factors responsible for generating such observed chaotic behavior, which was predicted previously in a three-level optically pumped laser system [153]. However, for two-level AOB systems, conditions for such mechanism of observing chaos will be too hard to fulfill in practice.

5.4.1   *Observation of period-doubling to chaos*

The experimental setup to observe chaos in the three-level atomic medium inside an optical cavity is the same as the one used for experimentally observing AOB, and instability as discussed in detail in Chapter 3 with reference to Fig.3.14. The only major difference is that in the experiment of demonstrating chaos, the EOM was not used for the cavity input field. The controlling experimental parameters in demonstrating chaos were intensities and frequency detunings of the cavity input laser beam and coupling laser beam, temperature of the rubidium vapor cell, and the cavity frequency detuning Δ_θ. To carry out experiments on chaos, experimental parameters were first adjusted so that the cavity output profile displayed the self-pulsing oscillation while the optical cavity was being scanned. The procedure and parameters of observing self-pulsing oscillation were described in section 5.3. Then, all other parameters were kept unchanged except Δ_c. The scan of the optical ring cavity was then stopped and it was frequency-locked to a cavity mode with the help of a third reference laser beam [47]. The experiment was done by measuring the cavity output profiles while decreasing Δ_c.

The experimental observations of the cavity transmission profiles with different coupling field detunings Δ_c are displayed in Fig.5.9. The remaining experimental parameters are $P_c = 19.7$ mW (before it enters the PBS), $\Delta_p = 0$, $\Delta_\theta = 0$, $T = 85$ ^{0}C, and $P_p = 16.6$ mW. Note that this value of P_p is at the entrance of the cavity. These parameters were kept unchanged during the experimental measurement. For large Δ_c (> 100 MHz), the system remains at the steady-state in the upper branch of the AOB curve, and hence the cavity output shows stable behavior, as in the case of a typical stable AOB output. As Δ_c is decreased near to 60 MHz, the cavity output shows an oscillatory behavior, or the system moves into the instability region as described in section 5.3. As Δ_c is reduced further the amplitude and period of oscillation change continuously [5,46,47]. When the value of Δ_c reaches about 20 MHz, half of the peaks change their amplitudes among the original peaks [Fig.5.9(b)] of the oscillatory pattern, and the oscillation period is said to be doubled or 'period-doubling' takes place. When Δ_c equals to 10 MHz, another set of oscillation peaks change their amplitude and the oscillation period becomes 'quadrupled', as shown in Fig.5.9(c). As the value of Δ_c is further reduced down to below 7 MHz [Fig.5.9(d)], the output oscillation loses its periodicity and becomes chaotic. As established previously, the linear dispersion and Kerr nonlinearity are both at their maximal values near such coupling frequency detuning (7 MHz) [25,33,47], which is a clear indication that the enhanced linear dispersion and nonlinearity in the three-level EIT system are responsible for the chaotic behavior here. These experimentally measured behaviors are in good agreement with the theoretical calculations discussed in the following subsection. The power spectra (that is another diagnostic tool to characterize chaos) corresponding to these time series are displayed in Fig.5.10. The features of spectrum change from well defined sharp spikes (usually observed for

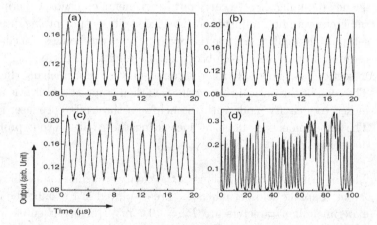

Fig. 5.9 Experimentally measured cavity output field as a funtion of time at four different values of coupling frequency detuning Δ_c at (a) 60 MHz, (b) 20 MHz, (c) 10 MHz, and (d) 7 MHz (reprinted from [47] with permission).

a periodic system, as shown in Fig.5.10(a)) to a broadband or diffused spectrum with random components (a typical spectrum for the chaotic system), shown in Fig.5.10(d), as Δ_c decreases. Figures 5.10(b) and 5.10(c) indicate the increase of the periodicity, i.e., more frequency components when the coupling (control) frequency detuning decreases. Thus, the route from "period-doubling to chaos" has been clearly established in the experiment, confirmed by the spectral analysis from the power spectrum plots of Fig.5.10 [5,47].

5.4.2 *Theoretical calculation*

A theoretical model with coupled density-matrix equations for a three-level atomic EIT system, together with field equations [46,47], can be used to study such chaotic behavior governing the current system established in the experimental observations. The dynamical equations governing the current system are the same ones as mentioned in Eq.(5.4), which were used for simulating the instabilities in subsection 5.3.2. As shown in the following, the simulated results match quite well with the experimentally observed phenomenon of reaching chaos through the period-doubling route [47]. This model used for simulating chaotic phenomenon in AOB of three-level EIT medium is surpassing the usual Lorentz model used for studying chaos in two-level atomic systems and lasers [152]. In this model the single-mode mean-field approximation has been employed, but the second electromag-

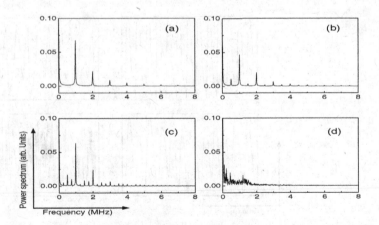

Fig. 5.10 Corresponding power spectra of cavity output shown in Fig.5.9. Coupling frequency detunings Δ_c are at (a) 60 MHz, (b) 20 MHz, (c) 10 MHz, and (d) 7 MHz, respectively, (reprinted from [47] with permission).

netic field, i.e., the coupling laser field, does appear through the atomic density-matrix equations. In this way the role played by the coupling laser field in controlling, and manipulating the system, as well as deciding the route taken by the system to reach chaotic regime, becomes quite important.

Dynamical behaviors of the cavity field can be obtained through the numerical solutions of Eq.(5.4). In Fig.5.11, plots of calculated cavity transmission intensity for four different values of Δ_c (as used in Fig.5.9) are shown. The initial conditions for the density matrix elements and the values of other parameters used to solve Eq.(5.4) are $z_p(0) = -1$, $z_c(0) = 0$, $x_{21} = x_{23} = x_{31} = 0 = y_{21} = y_{23} = y_{31} = 0$, $\alpha_1(0) = 6 \times 10^6$, $\alpha_2(0) = 0$, $\gamma_{cav} = 0.03$. Note that the value of parameter γ_{cav} selected here is towards the higher side of its theoretically defined value of $t_1^2/2$. This increase is because the partial transmittance of mirror M1 and the losses given by intracavity PBS are also included in this parameter. Also, some other parameters used in the calculation are: $\gamma_{21}/2\pi = \gamma_{23}/2\pi = 3 \times 10^6$ Hz, $\gamma_{31}/2\pi = 0.1 \times 10^6$ Hz, $\tau = 1.2$ ns, $t_1 = 0.17$ (corresponding to a intensity transmissivity of 3% for M_1), $P_c = 20$ mW, $P_p = 15$ mW, and $\Delta_p = 0$ [47]. For the coupling frequency detuning Δ_c to be 60 MHz, the cavity output exhibits a periodic oscillation, as shown in Fig.5.11(a). By decreasing Δ_c to about 20 MHz, the regular set of oscillation bifurcates into two alternating oscillations. In other words the period of oscillation becomes twice that of the original pe-

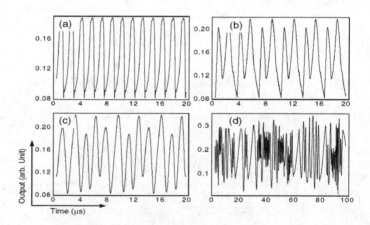

Fig. 5.11 Simulated cavity transmission profiles at four different values of the coupling frequency detunings Δ_c at (a) 60 MHz, (b) 20 MHz, (c) 10 MHz, and (d) 7 MHz, corresponding to the plots in Fig.5.9 (reprinted from [47] with permission).

riod and thus a period-doubling bifurcation [143,152] has taken place, which is clearly seen in Fig.5.11(b). By decreasing Δ_c further down to 10 MHz, the oscillation bifurcates again and the oscillation period becomes quadrupled or four times of the original time period as displayed in Fig.5.11(c). As Δ_c is decreased a little more and when it reaches a value to be less than 7 MHz, it is very difficult to find any regular periodicity in the oscillatory cavity transmission profile and the system is then said to be in the chaotic regime as shown in Fig. 5.11(d). So, the route to chaos in this EIT system is through period doubling [143,152] when Δ_c decreases from 60 MHz to around 7 MHz, which agrees with the experimental observations [47] shown in Fig.5.9.

To establish the chaotic nature of the experimentally obtained data, another diagnostic tool of determining Lyapunov exponent (LE) was employed in that work [47]. The LE quantifies the rate of divergence or convergence of phase space trajectories (originating from nearby initial points) in time for any system during its dynamical evolution. The sensitivity of system on initial conditions and the existence of 'attractor' can be investigated with the help of LE [143]. Phase space trajectories of a chaotic system diverge away from each other in an exponentially faster manner with a very little change in initial conditions [154]. To confirm chaos in the system, the largest value of LE needs to be calculated and that should be positive [47,154]. One can understand this by the following consideration. Let there be a tiny hypersphere of initial conditions in the phase space of the system. Due to the dynamical evolution, this hypersphere will be reshaped into a hyperellipsoid, meaning some of the axes will be elongated and others reduced. The axis having largest asymptotic expansion rate provides the most unstable flow direction of trajectories and can be determined by knowing the largest LE value [154]. Let there be two nearby initial points q_i and $q_i + s_i$ in the phase space. One can call s_i as a very small perturbation with respect to point q_i. After system evolves for a time t, the points q_i and $q_i + s_i$ will be represented by $g(q_i, t)$ and $g(q_i + s_i, t)$ and the new perturbation can be defined as [154]

$$s_f = g(q_i + s_i, t) - g(q_i, t) = Q_{q_i} g(q_i, t) \cdot s_i. \qquad (5.8)$$

In Eq.(5.8) the last term has been obtained by linearization of g. One can now obtain the mean exponential rate of divergence or convergence for the nearby trajectories by finding the value

$$\Lambda(q_i, s_i) = \lim_{t \to \infty} \frac{1}{t} \ln \frac{\|s_f\|}{\|s_i\|} = \lim_{t \to \infty} \frac{1}{t} \ln \|Q_{q_i} g(q_i, t) \cdot s_i\|. \qquad (5.9)$$

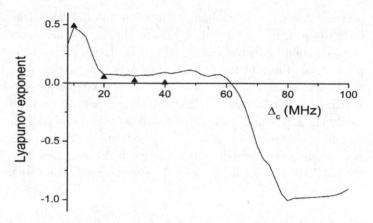

Fig. 5.12   Lyapunov exponents. Solid line gives theoretically simulated values and solid triangles are the numerical values calculated from experimental data (reprinted from [47] with permission).

Here the length of a vector s is represented by $||s||$. Clearly for $\Lambda(q_i, s_i) > 0$, trajectories corresponding to nearby points diverge exponentially. The limit given in Eq.(5.9) exists under very weak smoothness conditions, having finite values for almost all points belonging to initial hypersphere and thus provides the largest value of LE [154]. For our system, the value of largest LE has been calculated as a function of the coupling frequency detuning Δ_c, as displayed in Fig.5.12. In Fig.5.12, the solid line shows the theoretically calculated results. The region of Δ_c values where the calculated LE has negative value is a 'region of stability' for the system. This region is located at large values of Δ_c. The calculated LE value reaches very close to zero point when Δ_c is set around 60 MHz. At this point the system goes to the region with near zero LE and the cavity transmission field displays a regular oscillatory behavior. Due to certain computational difficulties in the numerical method, the LE values do not showing perfect zero values and some residual small positive values maintained to keep the stability for the computational calculations. As soon as Δ_c is lowered down below 20 MHz, there is a large jump in the computed LE value (LE becomes positive) and its value becomes 0.5 when Δ_c goes down below 7 MHz. When LE has a large positive value, the system clearly evolves in the chaotic regime. LE values are also computed from the experimentally observed time series, which are shown as solid triangles in Fig.5.12. There are good agreements between the experimentally measured LE values and theoretically calculated LE values without using any adjustable parameters [47].

In this experimental system with three-level atoms inside an optical cavity, variations of other parameters, e.g., probe frequency detuning, cavity detuning, probe power and coupling power, were also performed to search for the behaviors of instability and chaos, but the phenomenon of period-doubling could not be observed when the coupling detuning was kept large. One reason for not observing period-doubling to chaos, when the probe and coupling power are changed with large coupling (control) field frequency detuning, might be due to the limited output powers of the two diode lasers So, in order to obtain period-doubling route to chaos at higher frequency detunings, the output powers of the two lasers should be very high and the atomic number density should also be large (temperature of the atomic gas cell should be > 80 °C) [47], which modify the parametric positions where the maximal nonlinearity and dispersion occur [25,33]. It will be interesting to further explore larger parametric regions for instability and its route to chaos, especially the relations between the dynamical behaviors and the induced atomic coherence in this coherently prepared atomic system with enhanced nonlinearity and dispersion. Possibly, a four- or five-level system may be more suitable for this purpose because such systems have more flexibility to control nonlinearity and dispersion by means of several experimental parameters.

Chapter 6

Controlled All-optical Switching

6.1 All-optical Switching

An all-optical switch is a device that uses one optical signal to control another optical signal [155]. Usually in an all-optical switch, light is directed to go from one communication channel to another, with the help of another light, without undergoing any intermediate conversion to an electrical signal. In most current optical communication systems, optical signals are first converted to electronic signals, which can then be used for electronic triggering of any other light source to create (and code information onto) other optical signals. However, in the all-optical switching scenario, the signals will always remain as optical signals but interact with each other through specific medium, so one of the optical signal can be controlled by the another one directly via such interaction. In other words, an optical beam (or optical signal) passes through a medium and changes its optical property, which enables another light beam passing through the same medium to be controlled with the information coded by the first beam. The first light beam is therefore called the switching (or control) beam. The switching beam can control the intensity-dependent refractive index of the medium, such as the one associated with the third-order nonlinear properties of materials, via the cross-phase modulation as discussed in Chapter 1. In an all-optical switch, the response of the third-order Kerr nonlinearity can be in the range of picoseconds to femtoseconds time scale [156].

Typical characterizing properties of an optical signal include its phase, polarization, spatial location, and arrival time etc. The redirection of an optical signal can be achieved by changing any of the above characteristics by optically induced changes in a medium. A simple example of an all-optical switching device is the guided-wave Mach-Zehnder interferometer. This guided-wave Mach-Zehnder interferometer contains six different parts. Out of those, two are the input and output channels. The Y branch part of Mach-Zehnder interferometer bifurcates the input signal into equal components in the two intermediate channels [156]. The two signals propagate in the separate channels, almost similar in all respect and hence pick up equal phase shifts if there is no additional intense optical beam. Hence the recombination of two signals takes place in phase at the second Y junction. Thus the output signal is same as the input signal. If a nonlinear medium is kept in one arm of the interferometer, then the interaction of an intense control beam through it will produce an additional phase shift due to the generation of intensity-dependent refractive index. In this situation, at the output Y junction, signals in the two arms will not maintain the same phase. So, there is a partial destructive interference that reduces the transmitted signal. It is possible to have a very large phase shift induced in the arm containing nonlinear medium such that complete destructive interference occurs and no light through the second Y junction comes out. This leads to no signal at output and thus completing 'off' mode of the all-optical switching [156]. The change in refractive index of the arm containing nonlinear medium can be made to go from low to high value in time, when the illumination intensity varies in similar fashion. In this way the time dependent 'on' and 'off' of the output signal can be generated and an all-optical switching can be realized.

One of the most needed functions in all-optical communication and all-optical computation is to achieve effective and fast all-optical switching. The difficulty is to find systems with the right light-matter interactions necessary to make practical all-optical switches [157]. Practical all-optical switches will allow the circumvention of optical-electrical-optical conversions, which are the bottleneck for the high speed optical communication. Developing efficient optical switching was the main objective in the two-level AOB research thirty years ago (see [1,2] and references therein). Figure 6.1 illustrates the basic concepts of controlling light by light [157].

The AOB systems as well as other semiconductor bistable systems can be used as potential systems for all-optical signal processing and all-optical switching. The devices work as optical logic elements in which one light beam controls another. The electromagnetic radiation in optical regime can give subpicosecond switching times for optical logic elements. Since

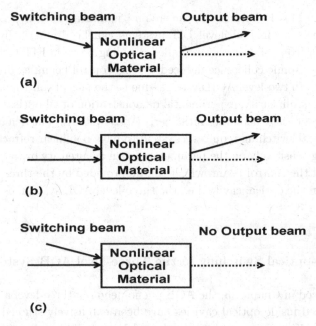

Fig. 6.1   Schematic illustration of all-optical switches (a) that changes the direction of output light, (b) that redirects light (c) an absorptive all-optical switch. In all three cases, the dotted arrow shows output beam when no switching beam is present (adopted from [157] with permission).

electrical charges do not participate in all-optical switching systems, there is no electromagnetic interference from the environment in such devices. One can have a variety of all-optical devices such as all-optical transistors, limiters, discriminators [2].

The theoretical model of all-optical switching using nonlinear phase shift [2] was given in 1980s and later on further investigated by using both second- and third-order nonlinearities. Those works involved couplers [158] based on ultrafast electronic nonlinearities and required large switching powers in the form of short pulses, hence no use for practical purposes. However, in network reconfigurations, fast switching devices are not so critical, so materials having slower response time, such as photorefractive, thermo-optic, optomechanical, liquid crystals, and molecular nonlinearities, can be used for optical switching devices [155].

Multilevel EIT systems are ideal candidates to realize all-optical switches

[159-161]. This is because the linear and nonlinear optical properties of the probe beam in multilevel EIT systems can be dramatically changed by another 'control' laser beam (called coupling beam in EIT terminology) due to the atomic coherence induced by this control beam, as discussed in Chapter 1. In two-level AOB systems, due to the lack of control parameters and small nonlinearity, experimental demonstration of all-optical switching has not been easily achieved in the early AOB systems. The demonstration of all-optical switching requires a system to have nonlinear refractive index depending sensitively on the parameters (such as intensity or frequency detuning) of the control beam, which can be provided by the three-level EIT medium in the optical cavity, i.e., the three-level AOB system, as discussed in Chapter 3.

6.2   All-optical Switching in the Three-level AOB System

As discussed in Chapter 3, the AOB phenomenon in three-level atomic systems placed inside optical cavities have been extensively studied in recent years by using improved theoretical models as well as well-controlled experiments. The three-level atomic systems can provide several advantages over the two-level system, especially on the controlled all-optical switching. The most important advantage, as already discussed several times in the book in the preceding chapters, is the induced atomic coherence in the three-level EIT system, which can greatly modify the absorption, dispersion, and nonlinearity of the probe laser beam by adjusting the control laser beam. Such modified linear and nonlinear optical properties for the cavity field can significantly reduce the cavity input threshold power for realizing optical bistability. In experiments it is easy to control the switching threshold value of the AOB with the help of intensity or frequency detuning of the coupling beam (as discussed in Chapter 3) and hence the width of the AOB hysteresis cycle can be controlled. Also, the experimental system to demonstrate AOB becomes much simpler when three-level atoms are used instead of two-level atoms as the intracavity medium, since the first-order Doppler effect can be canceled when a two-photon Doppler-free configuration is employed for the three-level system [24]. In this case, AOB can be studied in a simple experimental setup as discussed in Chapter 3. Note that in order to eliminate the first-order Doppler effect, the two laser beams have to co-propagate for the Λ-type configuration of the three-level atomic medium [24]. There are different ways to realize controlled switching in

AOB systems, some of which will be described in the following sections.

6.2.1 *Deterministic switching: controlled switching between bistable states*

Before demonstrating all-optical switching in three-level AOB systems, a simple controlled switching scheme will be described, which works for any AOB systems with clear hysteresis curves including the two-level AOB. The advantage of doing such experimental demonstration in a three-level AOB lies in the flexible control in the shape, width, and thresholds of the hysteresis loop, as demonstrated in Chapter 3. In this scheme, the controlling signal is coded on the cavity input beam, and the additional control beam is only used to control the shape of the hysteresis loop, so the deterministic switching can be easily achieved.

First, let's look at a typical hysteresis curve for AOB, as shown in Fig.6.2(a). By properly selecting the experimental parameters in the three-level AOB system as described in Chapter 3, an appropriate bistable curve with certain width for the bistable region can be chosen. The output cavity intensity will reside on the lower branch of this AOB hysteresis curve, when the input intensity is increased from initial zero or a lower value outside the AOB hysteresis curve region. As the cavity input intensity approaches the upper threshold value Y_2, it jumps to the upper branch of the AOB hysteresis curve. On the contrary, if the AOB system is initially on the upper branch of the hysteresis curve and the cavity input intensity is decreased continuously, the cavity output intensity maintains its level and remains on the upper branch until it reaches to a threshold value Y_1, then it jumps down to the lower branch of the hysteresis curve [161]. Hence it is possible to use such AOB property to implement an optical switch as described below. At the start of experiment, the input intensity is set at the value in the mid-region of the bistable curve such that the cavity output intensity is initially at the lower branch (point A) of the hysteresis curve. Then, an intensity pulse with its peak value higher than the threshold value Y_2 is applied to the cavity input laser beam, which brings the cavity output intensity to the upper branch of the hysteresis curve, i.e., point C in Fig.6.2(b). At the end of the short pulse when the input intensity returns to its initial value, the cavity output intensity stays on the upper branch at position B. By applying another negative pulse (which suddenly reduces the input intensity), the cavity output intensity comes down to the lower branch A (through

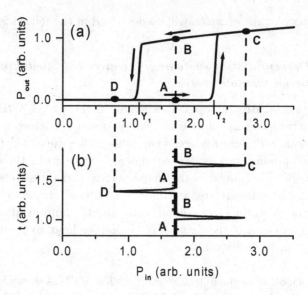

Fig. 6.2 Schematic diagram to illustrate the switching mechanism in the system. (a) A typical input-output intensity bistable curve. A, B, C, and D are output intensities for corresponding input intensities in (b), which shows a time sequence of the cavity input intensity (reprinted from [161] with permission).

point D) again. In this way an optical binary switch is accomplished. The two binary states are the low and high intensities of the cavity transmission output. This deterministically controlled optical switching was experimentally demonstrated and reported in Ref.[161].

The three-level Λ-type rubidium atomic system used for such switching experiment was the same as the one used in demonstrating the controlled AOB (see Fig.1.9(b)), where $F = 1$ (state $|1>$) and $F = 2$ (state $|3>$) states of $5S_{1/2}$ are the two lower states and $F' = 2$ (state $|2>$) of $5P_{1/2}$ serves as the upper state. The frequency of the control laser (or the coupling laser) beam is ω_c and it interacts with the transition between states $|3>$ and $|2>$. The frequency of the cavity field beam is ω_p, which interacts with states $|1>$ and $|2>$. The frequency detunings for the cavity and control beams are given by $\Delta_p = \omega_p - \omega_{12}$ and $\Delta_c = \omega_c - \omega_{23}$, respectively. Frequencies for atomic transitions from states $|1>$ to $|2>$ and from states $|2>$ to $|3>$ are given by parameters ω_{12} and ω_{23}, respectively.

In the experiment the initial input power going to the cavity was kept at 1.73 mW. The EOM (which was used to scan the input intensity in the AOB experiments) was applied with appropriate pulse voltage to generate the positive and negative pulses described above. The magnitudes of the positive and negative intensity pulses were chosen to be slightly higher than the half width of the bistable hysteresis curve. Initially the output intensity stayed at lower branch (state A). Applying the positive pulse through EOM (see Fig.6.3(a)), the output intensity was brought to the upper branch (state B in Fig.6.3(b)) of the hysteresis curve. The system stayed at the upper branch until by application of a negative pulse, which brought it down to the lower branch value (state A) again. The sharp spikes on the left of the output intensity square pulses at state B in Fig.6.3(b) are due to the slope in the upper branch of the bistable curve shown in Fig.6.2(a), since the output intensity at point C is higher than the intensity at point B. The states A and B are very stable and this switching action could provide an extinction ratio of about 20:1 [161]. Since the AOB hysteresis curve in such three-level atomic system can be well controlled, it is possible to have the threshold value very low and the hysteresis cycle narrow

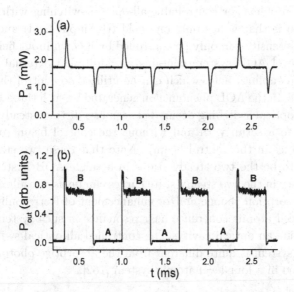

Fig. 6.3   Optical switching of cavity output intensity controlled by cavity input intensity. (a) Cavity input power as a function of time with positive and negative pulses, respectively. (b) Output steady-state intensities A and B controlled by the pulses in (a). The parameters are $\Delta_p = -76.9$ MHz, $\Delta_c = -19.3$ MHz, and $P_c = 9.5$ mW. The temperature of the rubidium cell is 68 ^{0}C (reprinted from [161] with permission).

for such AOB, which then reduces the required input magnitude of the pulses for realizing such switching process. The switching speed involved with such process in the experiment reported in Ref.[161] was limited to microsecond time scale. This limitation was imposed by the speed of the EOM (200 kHz maximum) used to generate intensity pulses. A much faster switching is possible by employing a faster EOM device. The limitation of switching time will then be determined by the cavity response time. The optical switching action observed in the above described experiment has a quite different physical mechanism from the all-optical switching schemes observed by manipulating control laser beam parameters [159,160]. In the later cases, the nonlinear optical processes in the atom-cavity system play a vital role for realizing the controlled all-optical switching of the cavity field by the control laser beam.

6.3 Controlling Light with Light in the Three-level AOB System

The crucial problem for controllable all-optical switching with a two-level AOB system is that it has only one field (the input field) and hence the cavity field intensity can only be controlled by its own input field intensity. The three-level AOB system contains an additional external field called controlling (coupling) field, which can be utilized to control the switching (probe) field. In the AOB phenomenon generated by such three-level atomic systems all-optical switching occurs between two distinct steady states corresponding to a setup with and without the control beam (or with two different values in the control beam). Note that these two states are not necessarily to be the two steady states in a single AOB hysteresis curve. The difference in the two steady-state curves can be as a result of combined effects of absorption change and/or enhancement of Kerr nonlinearity due to the induced atomic coherence near resonance in such system with EIT medium inside an optical cavity. The controlled all-optical switching arising in such system is quite different from the absorptive photon switching demonstrated in a four-level atomic system [162].

The three-level Λ-type rubidium atomic system used for demonstrating such all-optical switching was the same as the one used in the controlled AOB experiments (see Fig.1.9(b)). The details on atomic transition levels in rubidium atom and the corresponding transition frequencies, laser frequency detunings etc are discussed in section 6.2 above and in Chapter 3. The experimental setup is exactly the same as shown in Fig.3.14 and the details of this setup are provided there in the subsection 3.2.2 of Chapter 3. There was an additional EOM (designated as EOM1) placed just before the PBS in the cavity for the control beam (diode laser LD1). The EOM in the path of the 'probe' (switching beam) diode laser LD2 is now called EOM2.

The mechanism for realizing this intensity switching is based on how the cavity field intensity changes as the control field is turned on and off [159]. So, the first important task of the experiment was to measure the absorption and Kerr nonlinear index of refraction of the system. For absorption measurement the optical cavity was not utilized and the probe laser frequency was scanned so that Δ_p changed and the absorption in single-pass was measured to be approximately 5-6% when $P_c = 0$. This little amount of absorption can be explained due to the presence of saturation. For $P_c = 14.3$ mW and $\Delta_c = 7$ MHz, with a probe laser power of 120 μW, the absorption was measured to be 13-15% at $\Delta_p = 0$, where the operation of switch will take place. Next, the self-Kerr nonlinear index of refraction of this system was measured using the procedure described in subsection 1.5.4 of Chapter 1 [33]. Now the frequency scan of the switching (probe) beam was stopped and its frequency was set to the value so that $\Delta_p = 0$. The optical cavity was used at this stage and it was scanned from longer length to shorter length by applying a ramp voltage on the PZT on which one of the cavity mirrors was mounted. This was done for the above mentioned experimental parameters for both $P_c = 0$ and $P_c = 14.3$ mW. The transmission profiles of the optical cavity are shown in Fig.6.4. For $P_c = 0$ (curve a, which is the case for a two-level atomic system) the cavity transmission profile is narrow and apparently symmetric. Such transmission profile represents a small Kerr nonlinear index of $n_2 = -8.1 \times 10^{-9}$ cm^2/W, corresponding to a value in the two-level system. On the other hand, for $P_c = 14.3$ mW (curve b) the cavity transmission profile becomes asymmetric and broader, indicating a large Kerr nonlinear index of $n_2 = 2.5 \times 10^{-7}$ cm^2/W, as shown in subsection 1.5.4 of Chapter 1. The measured value of n_2 in this situation is smaller in comparison to the value reported in Ref.[33]. This is because the intensity of the cavity field used in this experiment was higher compared to the one used in Ref.[33], thus reducing the value of measured nonlinear index [34]. The sharp linear and the nonlinear dispersion effects near resonance [25,34] are responsible to induce the observed shift in resonance

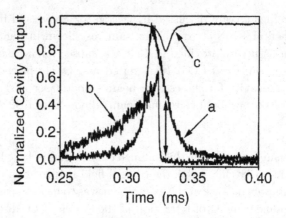

Fig. 6.4   Cavity transmission profiles by means of cavity-length scan. a, $\Delta_c = 7$ MHz, $P_p^{in} = 240$ μW, $P_c = 0$. b, $P_c = 14.3$ mW, with other parameters the same. c, Frequency of the reference laser beam used to lock the ring cavity (reprinted from [159] with permission).

peak of the transmission profile. The combined expression of the linear and nonlinear susceptibilities (χ) has been provided in Eq.(1.61) of Chapter 1, which clearly brings out the dependencies of susceptibility χ on parameters such as P_c ($= |E_c|^2$), P_p ($= |E_p|^2$), Δ_c, Δ_p etc.

A third laser (reference laser beam) was used to lock the optical ring cavity and curve c in Fig.6.4 shows its cavity resonance. With the help of the third laser (reference laser beam), cavity frequency detuning (Δ_θ) can be easily set to any desired value. After locking the optical cavity, the input cavity power was scanned using EOM2 for the two controlling-beam power values $P_c = 0$ and $P_c = 14.3$ mW, respectively, keeping $\Delta_p = 0$ and $\Delta_c = 7$ MHz fixed. The plots showing relationship between the input and output intensities of the optical cavity for the two cases, i.e., in the absence and presence of the control beam, are displayed in Fig.6.5(a) and Fig.6.5(b), respectively. In Fig.6.5(a), no AOB behavior is seen since the Kerr nonlinear coefficient is small in this two-level case. AOB is clearly visible in Fig.6.5(b) as the Kerr nonlinearity is large in this situation because the control beam has a non zero power (with EIT-enhanced Kerr nonlinearity [34]). The basic difference between these two steady-state curves can be attributed to the joint effects of the near-resonant absorption and the enhancement of near-resonant Kerr nonlinearity. With these two input-output intensity plots, the cavity input power was then set at $P_P^{in} = 240$ μW approximately (see

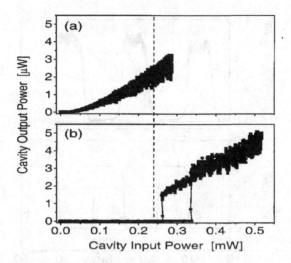

Fig. 6.5 Cavity transmission power versus cavity input power. These are the steady-state curves of the system for (a) $P_c = 0$ and (b) $P_c = 14.3$ mW, with $\Delta_c = 7$ MHz, $\Delta_p = 0$, and $P_p^{in} = 240$ mW. These curves were taken with cavity input intensity scanning up and then down (reprinted from [159] with permission).

the dotted vertical line in Fig.6.5) and the control laser beam was modulated with EOM1 using a square waveform generator. The application of square wave modulation in the control laser beam makes it turn on and off (Fig.6.6, bottom) and thus the transmitted power out of the cavity is also switched between two steady-state values (Fig.6.6, top). In this case an all-optical switching action is performed by the cavity field controlled by the modulated control beam with a switching ratio of 20:1 or better [159]. The physical mechanism behind above described all-optical switching can be easily understood from Figs.6.4 and 6.5. The transmission curves a and b of the optical cavity depicted in Fig.6.4 are for two different power levels of control laser, e.g., $P_c = 0$ and $P_c = 14.3$ mW, respectively. When P_c switches between these two values (i.e., on and off) the cavity transmission also switches between two steady-state values as marked by the arrows. Note that the optical cavity was locked at a cavity detuning of $\Delta_\theta = 4.6$ MHz by the transmission peak of curve c. In Fig.6.5, in the absence of the control beam, e.g., $P_c = 0$, the transmitted intensity (power) out of cavity increases uniformly with the cavity input power, which is because of the saturation effect. Both absorption (about 5–6% per pass) and nonlinearity are small and AOB can not be created under such (two-level) experimental conditions. Turning on the control beam to a power

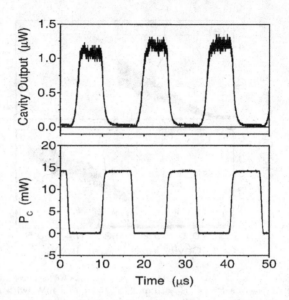

Fig. 6.6   Optical switching of the cavity field controlled by the controlling beam. Top, ON and OFF states of the cavity field. Bottom, modulated controlling power(reprinted from [159] with permission).

of $P_c = 14.3$ mW, both absorption (now becomes 13-15%) and enhanced Kerr nonlinearity jump to much higher values to produce AOB. The upper branch of the AOB exists when $P_P^{in} = 260$ μW approximately. The switching experiment works here when $P_P^{in} = 240$ μW just before starting of the bistable region. This switching action described above can be realized in a wide range of experimental parameters since both absorption and nonlinearity can be used for changing the steady-state behaviors of such composite system comprising of optical cavity filled with multilevel atomic vapor [34].

In such all-optical switching scheme, the optical information (binary bits of 'on' and 'off') is encoded on the control beam, which is then transferred all-optically via this atom-cavity system into the cavity field. The speed of switching in the experiment was limited to a few microseconds as the EOM1, which is responsible to switch the controlling power, could only operate up to 200 kHz using a sinusoidal waveform [34]. However, such experimental demonstration could be very useful in further understanding the relative roles played by the absorption and enhanced nonlinearity in multilevel EIT system to achieve all-optical switches at ultra low intensity levels for all-optical communications and information processing [96,97].

6.4 Controlling Cavity Output with Coupling Beam Frequency

In the above experiment, the control beam power has been turned on and off to provide modulation to the cavity output field. Actually, as shown in Fig.1.17 of Chapter 1 [33,34], self-Kerr nonlinear index seen by the cavity field can change significantly (by two-orders of magnitude or even sign) when the frequency detuning of the control (coupling) beam changes by a small amount (tens of MHz). This property has been used to construct a different type of all-optical switching scheme as the one described in the last section [160]. In this scheme, changing the frequency of the control laser beam by 24 MHz, the cavity-field intensity can switch with a switching time of few microseconds and switching ratio better than 30:1 [160]. In general, switching laser frequency between two values can be much faster and with less energy consumption than switching laser power, so such constructed all-optical switching can be more efficient. Also, such controlled frequency-to-intensity all-optical switching can better exploit one of the unique properties of the EIT medium, i.e., a largely enhanced nonlinearity near EIT resonance with a very sensitive frequency dependence for the control field (Fig.1.17) [33,34].

The experimental setup used in demonstrating this frequency-to-intensity all-optical switching is basically the same as the one described in section 6.3 above [160]. Also, since the basic mechanism of such controlled all-optical switching is also based on the cavity transmission characteristics, especially the absorption and enhanced Kerr nonlinearity, one needs to measure these properties first. The first measurement was done to obtain the absorption spectrum of the switching field by scanning Δ_p for fixed cavity input power $P_p^{in} = 390 \ \mu W$ and control laser power $P_c = 14.3 \ mW$. The control laser frequency detuning Δ_c, which controls switching, was chosen to move back and forth between the values of 111 and 135 MHz, where the atomic absorption for these two frequencies were about the same and small. The realization of switching for the frequency detuning (Δ_c) of the control laser beam between $\Delta_c = 111$ and 135 MHz, was obtained through the phase modulation of the control laser beam with an electro-optical modulator EOM1 placed in the feedback path of the diode laser [160]. The square-wave modula-

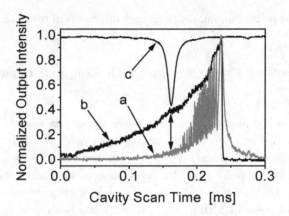

Fig. 6.7   Cavity transmission profiles via cavity length scan.   Curve a is for $\Delta_c =$ $\Delta_p - 12$ MHz with $\Delta_p = 123$ MHz, $P_p^{in} = 0.39$ mW, and $P_c = 14.3$ mW; curve b is for $\Delta_c = \Delta_p + 12$ MHz with other parameters the same; curve c is the frequency of the reference laser beam used to lock the ring cavity (reprinted from [160] with permission).

tion was applied to EOM1 with the frequency of 58 kHz and a rise time of < 1 μs. The input cavity field frequency was locked to $\Delta_p = 123$ MHz, which was midway between the two alternating frequencies of the control laser beam and acted as the operating point for switching [160]. The cavity was scanned from longer to shorter length at a rate of 0.903 μm/ms by a ramp voltage on the PZT, when the frequency detuning of the control beam was set at $\Delta_c = 135$ MHz (Δ_p+12 MHz) and $\Delta_c = 111$ MHz (Δ_p-12 MHz), respectively. In Fig.6.7 the output profiles of the cavity transmission intensity for the two different values of control beam detunings are shown. For $\Delta_c = 111$ MHz (curve a), the cavity transmission profile is not so broad, which is due to a lower magnitude of Kerr nonlinear index of refraction with a typical value of $n_2 = -0.8 \times 10^{-7}$ cm^2/W. On the left hand side of the transmission peak there are some oscillations, which represent the presence of optical instability in such system as discussed in Chapter 5. When the control beam detuning goes to $\Delta_c = 135$ MHz, the cavity output profile becomes highly asymmetric as shown in curve b, due to the increased nonlinear index of refraction ($n_2 = 6 \times 10^{-7}$ cm^2/W), as described in subsection 1.5.4 of Chapter 1 [33,160]. In this figure, curve c represents the inverted cavity output of a frequency-locked reference laser beam. The reference beam was detuned far from atomic resonances as the cavity length was scanned. The reference beam travels through the cavity in opposite direction with respect to both the control and probe beams [160]. An op-

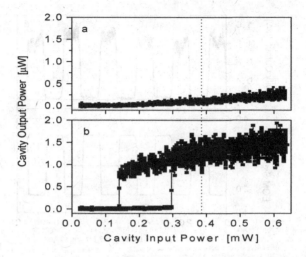

Fig. 6.8 Cavity input-output power curves. These are the steady-state curves of the system for (a) $n_2 = -0.8 \times 10^{-7}$ cm^2/W ($\Delta_c = \Delta_p - 12$ MHz) and (b) $n_2 = 6 \times 10^{-7}$ cm^2/W ($\Delta_c = \Delta_p + 12$ MHz), respectively. The cavity input power of the switching beam is set at 0.39 mW with all other parameters the same as in Fig.6.7 (reprinted from [160] with permission).

timum condition of the optical cavity detuning was chosen to operate this controlled switching to have a good switching ratios and to be away from the instability region of the system, which is indicated by the double arrow in Fig.6.7. The frequency locking of the optical ring cavity at the chosen value ($\Delta_\theta = 65$ MHz) was accomplished on the cavity transmission peak of the reference laser beam. The cavity input-output curves were recorded by scanning the cavity input power using an EOM for the same two frequency detunings of the control laser beam, respectively, as shown in Fig.6.8. The AOB observed is displayed in Fig.6.8(b) when the nonlinear index n_2 has a large value, i.e., $n_2 = 6 \times 10^{-7}$ cm^2/W under the condition $\Delta_c = 135$ MHz with a very low up-switching threshold power corresponding to a cavity input power of 0.3 mW. On the other hand, for $\Delta_c = 111$ MHz, the nonlinear refractive index is low with a negative sign, i.e., $n_2 = -0.8 \times 10^{-7}$ cm^2/W. Hence, the threshold for AOB increases drastically (or does not exist) and it could not be observed in the experiments with the available input power of 0.6 mW as displayed in Fig.6.8(a). When the cavity input power is fixed at $P_p^{in} = 0.39$ mW and the control laser frequency detuning is allowed to switch between values of $\Delta_c = \Delta_p + 12$ MHz and $\Delta_c = \Delta_p - 12$ MHz (Fig. 6.9(b)) with EOM1, then the output intensity of the optical cavity is forced

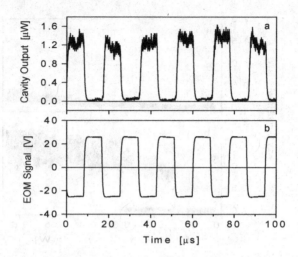

Fig. 6.9 Optical switching of the cavity field controlled by the controlling beam. (a) On and off states of the cavity field controlled by the voltage applied on EOM1, shown on (b), for two frequency detunings of $\Delta_c = 111$ MHz (upper voltage level) and $\Delta_c = 135$ MHz (lower voltage level), respectively(reprinted from [160] with permission).

to switch between two different steady-state values on two different steady-state curves (Fig.6.9(a)). Quantitatively, the cavity output peak power of the 'on' state is about 1.2 μW (on the average), and the power of the 'off' state is less than 0.034 μW (on the average) giving the switching ratio of better than 30:1 [160]. The all-optical switching mechanism in this composite system of optical cavity and three-level EIT medium can be understood with the help of Figs.6.7 and 6.8 in a straightforward manner. The cavity transmission profiles represented by curves (a) and (b) correspond to two different Δ_c values as shown in Fig.6.7. In order to demonstrate the switching phenomenon, the optical cavity has been frequency-locked at a frequency that corresponds to $\Delta_\theta = 65$ MHz, i.e., the resonant position of curve c. Then Δ_c was allowed to switch between the two values mentioned above, and consequently the cavity transmission intensity is also driven to the two different steady-state values [160], as shown in Fig.6.7 by a double arrow on curve a and curve b at the resonant position of curve c. For the selected parameters in the experiment: Δ_p, I_c, P_p^{in}, and Δ_c, the cavity detuning of $\Delta_\theta = 65$ MHz gave the maximum switching ratio [160]. When the optical cavity frequency was set nearer to its resonance ($\Delta_\theta < 65$ MHz), the system did not remain stable. The switching action so observed in the experiments could be interpreted as frequency-to-amplitude signal conversion because frequency modulation of one laser beam causes the intensity modulation of the other laser beam. These all-optical frequency-to-amplitude conversion could be of prime importance in all-optical communication and all-optical information processing.

One of the advantages of such all-optical switching is its efficiency as it requires to simply change the frequency detuning of the control laser beam back and forth by a very little amount, e.g., 24 MHz, which provides a switching ratio of better than 30:1 for the cavity transmitted intensity. The speed of switching (which is about a few microseconds) here is due to the limitation of the driver for EOM1, which is about 200 kHz maximum. It might also be feasible to demonstrate switching of the cavity transmission intensity with an even smaller change of the frequency detuning of the control laser beam at lower power of the ingoing cavity field. This is based on the fact that during measurements of the Kerr nonlinear index [33], a change in Δ_c by 7 MHz causes n_2 to change from a maximum value of 7×10^{-6} cm^2/W to zero at near resonance with a cavity input intensity as low as 10 μW approximately (see Fig.1.17). In order to show this experimentally with a smaller change of the control beam frequency detuning, better frequency stability of the lasers and the optical cavity are required. For the experiments discussed above the frequency detunings (Δ_p and Δ_c) of the two laser beams were selected with the considerations of the stability and repeatability of the switching phenomenon [160].

Absorptive photon switching in a four-state N-type atomic system by quantum interference was proposed in [163]. The system undergoes two-photon absorption but the single-photon absorption does not occur due to EIT condition. The optical switch in this situation is realized by absorption of light at a particular frequency due to the presence of an optical pulse at some other frequency. This switch is very delicate as it works on the quantum interference effect, which is very tenuous. In principle, this switch can work under single-photon pulse. This absorptive photon switching was demonstrated in experiments utilizing a cold four-level atomic system [162], but the switching ratio has been low in such absorptive switching schemes.

Electrical switches are extremely useful in many current applications. It enables to turn appliances 'on' and 'off' electronically and direct electronic signal-streams around a network. All-optical switches can in principle fulfill the same functions as all-electronic switches in optical applications, e.g., direct signal-streams around optical networks or serve as building blocks for optical computers. An area for which all-optical switches can be very important is optical communications, because nowadays most long-distance telephone and internet communications are carried on optical fibers but in most cases, signals have to be turned from light into electricity for any controls and modulations, as well as readouts. The processes of converting signals from light to electricity and back use extra power (and generate extra errors) that can be expensive and slow, as well as limiting the bandwidth, if the conversions have to be done many times in a row. The efficiency and bandwidth of optical communications can be increased dramatically if all the devices for optical signal processing and networks can be designed to perform all optically.

Chapter 7

Dynamical Hysteresis and Noise-induced Behaviors

7.1 Dynamical Hysteresis Cycle

In the previous chapters, all the studied phenomena have been achieved by keeping experimental parameters either fixed or tuned (scanned) adiabatically. Although the system becomes unstable in the cases of instability and chaos (Chapter 5), if the experimental parameters have changed. In other words, the system can always follow the changes of the experimental parameters, such as the adiabatic scanning of the cavity input intensity to observe AOB in Chapter 3. An interesting question to ask is what happens when the experimental parameters change faster than the system can respond. In this chapter, we will discuss few interesting effects that are caused by either a non-adiabatic change of experimental parameter, i.e., the intensity of cavity input field [164,165], or added external noises in the system. The typical AOB curve is a hysteresis cycle (HC) in the input-output intensity plot. Besides AOB, the HC behavior is also exhibited in magnetic, optical, electronic, mechanical, chemical, and biological systems. The appearance of such HC is an indication of some kind of 'memory' effect in the system [118]. The phenomena of HC behaviors are quite interesting from both fundamental physical as well as mathematical point of views. Also, such HC can be quite important in designing all-optical switches (as discussed in Chapter 6 above) and certain memory devices [2]. The typical HCs (or the input-output field intensity plot) in optical bistable systems of atoms (or AOB systems) and semiconductors are obtained when the cavity input intensity is changed adiabatically from low to high values and back, as demonstrated in Chapters 2 and 3 [1,2]. The bistable systems normally exhibit static hysteresis cycle (static HC) with a nonzero area in the limit of zero sweeping rate of the cavity input intensity (i.e., under the adiabatic

175

limit) [1,2] as the cases discussed in previous chapters. However, when the
variation (or sweeping) of the cavity input intensity is not adiabatic but
with a large sweeping frequency Ω, then dynamical hysteresis cycle (dynamical HC) behaviors appear in the system's input-output characteristics
[164,165]. There are considerable amount of changes in the shape and area
of the dynamical HC in comparison to the typical static HC. Dynamical
HC behaviors appear because the non-adiabatic variation of the input field
introduces a delay in the transition for the cavity field to jump from the
lower state to the upper state of the HC, i.e., the bistable system can not
follow the rapid change of the experimental parameter. The size of the dynamical HC in terms of its width or area enclosed is obtained when a plot
of cavity output field is made with respect to the cavity input field, which
is varying periodically in time with a sweep rate parameter. Such a plot is
called the input-output curve. Increasing the sweep rate usually results in
a larger hysteresis cycles because the bifurcation point is delayed [164,165].
In other words the acquisition of additional area in dynamical HC over the
static HC (as discussed thoroughly in terms of controlling its width and
shape in the Chapter 3) is related to the instability of the system, and
hence the system is unable to relax completely and the bifurcation point is
delayed [164,165].

7.1.1   *Theoretical model*

The static HC characteristics of AOB have been well established over the
years (see Chapters 2 and 3 on the discussions of static HC [1,2]) but the
dynamical HC behaviors have not been well studied in AOB systems, especially not much experiments done in such interesting topic. The dynamical
HC is also important in magnetic and optical switching devices. In such
devices the area of the HC gives the estimation of power dissipation by
repetitive switching at a frequency Ω [165]. Note that in describing static
HC in AOB, the density-matrix equations of the atomic system and the
Maxwell's equation for the circulating cavity field along with boundary
conditions are used (see Chapters 2 and 3 for detail discussions). However,
the phenomenon of AOB can also be described by a simplified phenomenological model of quartic potential well (i.e., a double well model), which
can be used to easily incorporate dynamical HC behavior. In the following, we adopt this simple approach to qualitatively describe the dynamical
hysteresis behaviors in the three-level AOB system. This model essentially
describes one-dimensional theory of dynamical hysteresis, which predicts

that the shift of intensity switching points and the area of HC scale as the two-third power of the sweeping frequency [164,165]. These predictions match well with the experimental results involving a bistable semiconductor laser [164,166]. In such bistable system, the dynamical hysteresis area was calculated analytically [164,165] using the following model equation incorporating the Kerr nonlinear term

$$\frac{dx}{dt} = px - qx^3 + G(t), \tag{7.1}$$

in which p and q are certain constants related to system parameters. The control parameter of the system is the external driving field $G(t) = E\sin(\Omega t)$. The value of E is sufficiently large so that the system can go past the turning points repeatedly. The system described by Eq.(7.1) gives the static HC of finite area in the limiting condition of zero-frequency external driving field [165]. Equation (7.1) can be physically justified by considering a quartic potential well for the longitudinal mode bistability of a semiconductor laser [167]. This equation also describes a system of particle in a quartic double-well potential with a periodic driving force G(t). The area $A_{hys}(E,\Omega)$ of the dynamical hysteresis loop in the plane $(x(t), E\sin(\Omega t))$ goes by the scaling law $A_{hys}(E,\Omega) - A(0) \propto \Omega^{2/3}$ under the limiting condition of $\Omega \to 0$. The area of the static hysteresis loop is given by $A(0)$. It was shown that this scaling law matches quite well with the experimental measurements on a bistable semiconductor laser system [164]. For small Ω, one can have a general scaling law for the area of dynamical HC given by $A(\Omega) \sim \Omega^\beta$, such that the exponent $\beta \to 1$ as $\Omega \to 0$ but otherwise $\beta < 1$ [164,165]. For larger Ω the scaling law becomes $A(\Omega) \sim 1/\Omega$ as $\Omega \to \infty$. In the following we present the derivations for the above mentioned expressions. For this purpose we closely follow Ref.[164].

If the control parameter G is changed adiabatically, then the stationary solution of Eq.(7.1) is given by the cubic polynomial $S(x) = -px + qx^3$ on the stable branches up to the switching thresholds (i.e., limit points S_1, S_2), as shown in Fig.7.1. The coordinates of the limit points are as follows:

$$S_1 : [x = x_0 = -\left(\frac{p}{3q}\right)^{1/2}, G = G_0 = \left(\frac{4p^3}{27q}\right)^{1/2}],$$
$$S_2 : [x = -x_0, G = -G_0]. \tag{7.2}$$

As soon as the system is near the limit point S_1 it jumps to the upper stable branch. However, increase of Ω causes limit point to be delayed and hence the area of hysteresis loop increases. It is possible to expand Eq.(7.1)

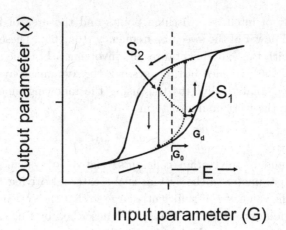

Input parameter (G)

Fig. 7.1   Static hysteresis loop (inner curve) for $G(t) = E\sin(\Omega t)$, with $\Omega \sim 0$, and the dotted curve (S-shaped) depicts theoretical bistability. Outer curve (dark solid line) is dynamical hysteresis loop for $\Omega > 0$ (adopted from [164] with permission).

about the limit point x_0 by setting $x = x_0 + z$ and $G(t) = G_0 + B\sin(\Omega t)$ to get a equation [164]

$$\frac{dz}{dt} = (\sqrt{3pq})z^2 + B\sin(\Omega t), \qquad (7.3)$$

where $B^2 = E^2 - G_0^2$ if the modulation frequency is small. By using a transformation $z(t) = [dw(t)/dt]/(\sqrt{3pq})w(t)$ and assuming $\sin(\Omega t) \approx \Omega t$, under the limiting condition $\Omega \to 0$, it is possible to rewrite Eq.(7.3) into a Mathieu equation

$$\frac{d^2w(\tau)}{d\tau^2} + \tau\Lambda^2 w(\tau) = 0, \qquad (7.4)$$

in which $\Lambda^2 = 1/3pq\Omega^2 B^2$ and $\tau = E(\sqrt{3pq})\Omega t$. The solution of Eq.(7.4) can be written by making use of Airy functions [164]

$$z(t) = \Lambda^{2/3} B\Omega \frac{Ai'(-\Lambda^{2/3}\sqrt{3pq}B\Omega t)}{Ai(-\Lambda^{2/3}\sqrt{3pq}B\Omega t)}. \qquad (7.5)$$

Here Λ is a parameter of Airy function. The delayed value of control variable defined by G_d is depicted in Fig.7.1. This value can be calculated by equating the derivative of the Airy function to zero. The estimated delayed critical value of the control parameter under small sweeping frequencies in comparison to the curvature of one of the unperturbed potential wells is [164]

$$G_d = G_0 + L_1[\Omega^2(E^2 - G_0^2)]^{1/3}, \qquad (7.6)$$

in which parameter L_1 is a function of p, q and can be estimated using other properties of the Airy function. Clearly, G_d varies as $\Omega^{2/3}$. The area of the hysteresis loop is given by

$$A_{hys}(E, \Omega) = A_{hys}(E, 0) + L_2[\Omega^2(E^2 - G_0^2)]^{1/3}. \tag{7.7}$$

The parameter L_2 related to zero of the Airy function is a constant (approximately). To obtain a well defined switching, the modulation must exceed G_d by a certain amount given by $E_d = \beta G_d$ and β should be larger than 1. This condition along with Eq.(7.6) self consistently provides the expression [164]

$$E_d = \beta G_0 + \beta L_1[\Omega^2(E_d^2 - G_0^2)]^{1/3}. \tag{7.8}$$

Eq.(7.8) can be solved to obtain the value of the modulation amplitude required for repetitive switching

$$E_d(\Omega) \sim E_d(0) + \beta L_1[\Omega^2(E_d^2(0) - G_0^2)]^{1/3}. \tag{7.9}$$

Clearly, the amplitude varies as $\Omega^{2/3}$. Since the input power is proportional to square of amplitude, so the input power required for repetitive switching varies as $\Omega^{2/3}$ in the leading order [164].

In a recent experiment dynamical HC has been demonstrated in warm rubidium atomic vapor using resonantly enhanced Raman generation without an optical cavity [168]. In yet another experiment [45], investigation of the dynamical HC in an AOB system, comprising of rubidium atomic vapor in a three-level Λ-type configuration (see Fig.1.9(b)) contained inside an optical ring cavity, has also been carried out with a much broader range of sweeping frequency Ω for the input intensity. This experimental system exhibits controllable AOB with flexible experimental parameters, as discussed in Chapter 3.

In view of these recent experimental developments in dynamical HC using three-level EIT systems, a physical interpretation of Eq.(7.1) can be provided for a nonlinear polarization model describing dispersive (refractive) optical bistability [169]. The cavity frequency detuning and third-order Kerr nonlinearity of the system can be directly related to the parameters p and q in Eq.(7.1), respectively [169]. The parameter q in the case of the three-level EIT system in Λ-type configuration of its levels can be written as (see Eqs.(1.61) and (1.62) in Chapter 1)

$$q \sim Re[\frac{-iN|\mu_{21}|^4}{3\hbar^3} \frac{F + F^*}{(2\gamma + \gamma_{21})F|F|^2}]. \tag{7.10}$$

Here γ_{21}, and γ_{31} represent the spontaneous radiative decay rates of the excited state $|2>$ to the ground states $|1>$ and $|3>$, respectively (see Fig.1.9(b)); γ_{31} is the nonradiative decay rate between the two ground states and $\gamma = (\gamma_{21} + \gamma_{23} + \gamma_{31})/2$. N is the atomic number density and μ_{21} is the transition diploe-matrix element between states $|1>$ and $|2>$; F is defined after Eq.(1.61). It has been well established that for such a three-level EIT system the parameters, Δ_C, N, and optical cavity detuning etc, can control its absorption, dispersion, and nonlinearity owing to induced coherence or quantum interference (as discussed in Chapter 1) [24,25,33]. Controls of the static AOB in such experimental system has already been demonstrated [42] and discussed in Chapter 3. The experimental demonstration of the controlled dynamical HC by varying sweeping frequency (Ω) of the cavity input intensity will be discussed in the following.

7.1.2 *Experimental results*

The experimental arrangements, including atomic medium, optical ring cavity, as well as the coupling and probe lasers, are exactly the same as the one used in the demonstration of controlling three-level AOB discussed in Chapter 3. The probe beam (ω_P) interacts with the atomic transition (ω_{12}): $5^2S_{1/2}(F=1)$ (level $|1>$) to $5^2P_{1/2}(F'=2)$ (level $|2>$) of ^{87}Rb atom, and circulates in the cavity. A sinusoidal modulation is superimposed on this field before it enters into cavity by an EOM. The control (coupling) beam interacts with the atomic transition (ω_{23}) : $5^2S_{1/2}(F=2)$ (level $|3>$) to $5^2P_{1/2}(F'=2)$ (level $|2>$) of ^{87}Rb atom, without circulating in the cavity.

As discussed in Chapter 3, to observe AOB from this composite atom-cavity system (described in Fig.3.14), the probe laser frequency was tuned near the preselected atomic transition (ω_{12}). After that the control laser frequency was set to another transition (ω_{23}) of the Λ-type three-level atomic system so that the EIT condition could be achieved. After locking the optical cavity to the reference laser frequency, the EOM modulated the intensity of the ingoing cavity field with a triangular shaped driving pulses. This allowed observations of output intensity from the cavity as a function of input intensity and thus HC curves could be monitored for different sweeping rates of the input intensity [45]. Parameters Δ_P or/and Δ_C should be nonzero in order to have HC in this AOB system. When a very low sweeping rate ($\Omega \sim 100$ Hz) was used for scanning the input cavity intensity, the HC could be observed with a moderate control beam power. At

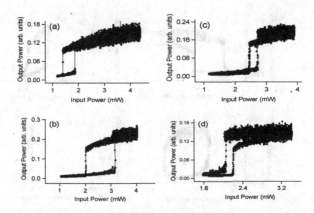

Fig. 7.2 The input-output intensity characteristics of the optical cavity field for different rates of triangular scan. The parametric conditions are $P_C = 14$ mW, $\Delta_C = 80$ MHz, $\Delta_P = 20$ MHz, $\Delta_\theta = 50$ MHz, $T = 70\,^{0}$C. Curves (a), (b), (c) and (d) are for $\Omega = 150$ Hz, 1.5 kHz, 2.5 kHz, and 3.0 kHz, respectively (reprinted from [45] with permission).

this sweeping rate, the system could follow the change of the cavity input intensity, and therefore static HC was observed, as shown in Fig.7.2(a). As discussed in Chapter 3, for small values of Δ_C, absorptive-type AOB could be observed but as Δ_C or Δ_P increased, dispersive-type AOB should be observed in the system [42]. By increasing the sweeping rate for the cavity input field but keeping all other parameters same, a drastic change in the shape of the HC was observed. Figure 7.2 displays dynamical HCs from the three-level Λ-type rubidium atoms under the experimental parametric conditions: $P_C = 14$ mW, $\Delta_C = 80$ MHz, $\Delta_P = 20$ MHz, cavity detuning $\Delta_\theta = 50$ MHz, and $T = 70\,^{0}$C. Plots (a), (b), (c), and (d) in this figure are for the sweeping rates $\Omega = 150$ Hz, 1.5 kHz, 2.5 kHz, and 3.0 kHz, respectively. For the chosen parametric conditions of the experiment, the AOB curves shown in Fig.7.2(a) and Fig.7.2(b) are typical more towards dispersive (refractive) type AOB. Under the adiabatic (or near adiabatic) condition of the sweeping rates, i.e., $\Omega \sim 100$ to 200 Hz, it was observed that the area of the HC does not change much. As Ω increases to 1.5 kHz, the area of the HC increases (Fig.7.2(b)) since the limit point (Fig.7.1) is delayed due to the increased sweeping rate. This implies that with an increased Ω, the system switches to the upper branch of AOB hysteresis curve in a delayed manner [45]. The scaling law of $A(\Omega) \propto \Omega^{2/3}$ (within the experimental uncertainties) could be observed in the HC curves for the values of Ω between 0.5 kHz to 1.5 kHz in consistent with the discussion above in the theoretical model and in Refs.[164,165]. By increasing the sweeping

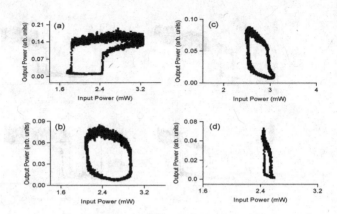

Fig. 7.3   Same as Fig.7.2 but curves (a), (b), (c), and (d) are for $\Omega = 5.0$ kHz, 10.0 kHz, 15.0 kHz, and 30.0 kHz, respectively (reprinted from [45] with permission).

rate further (say 2.5 kHz), the area of dynamical HC decreases in Fig.7.2(c) in comparison to Fig.7.2(b). When Ω goes to a further higher value (3.0 kHz), there is a drastic change in the shape of HC as shown in Fig.7.2(d). At this stage the HC area increases with an increased Ω, which is quite clear from Fig.7.3(a). By continuously increasing the sweeping rate Ω, the shape and area of the dynamical HC continue to modify (Fig.7.3(b)) in very unusual and interesting manner. It is easy to observe in Figs.7.3(c) ($\Omega = 15.0$ kHz) and 7.3(d) ($\Omega = 30.0$ kHz) that the hysteresis loop area decreases by the asymptotic scaling law of $A(\Omega) \sim 1/\Omega$ as $\Omega \to \infty$. This indicates that the dynamical HC area asymptotically approaches to zero value at very large sweeping rates [164,165]. This interesting behavior of dynamical HC is due to the dynamical cutoff stemming from the imbalance between the sweeping time period ($\sim 1/\Omega$) and the mean dwell time in the one state [45,170]. In the bistable system like AOB considered here, HC is observed because the system stays in one of the two metastable minima of the quartic potential well. If the sweeping rate (Ω) is high enough, the AOB system does not have enough time to remain in either of the metastable states, so the area of HC diminishes. The scaling laws for the area of dynamical HC work very well, at least qualitatively under the two limiting conditions of sweeping rates [45]. At intermediate Ω values, the behaviors of dynamical HC in terms of its area and shape could not be described by such simplified model as presented in Eq.(7.1). Hence, dynamical HC from a three-level AOB system exhibits very interesting behaviors when its shape and area are studied by varying the sweeping frequency of the cavity input field. To understand such dynamical HC phenomena completely, better theoretical modeling is needed in view of these experimentally observed results. Dynamical behaviors in double-well (two-state) systems, as presented here are interesting physical, as well as mathematical problems to study, since they have provided the ideal platform for investigating more complicated phenomenon in nature [48,49], which will be discussed in sections 7.2 and 7.3 in this Chapter. The experiments on dynamical HC in such simple atomic system could help to control hysteresis behaviors for potential applications in all-optical switches, optical transistors, and optical memory devices. It is quite likely to realize the dynamical optical bistability in condensed matter systems [171,172] which are more practical for applications as all-optical switching devices.

7.2 Stochastic Resonance in Atomic Optical Bistability

The phenomenon of stochastic resonance (SR) occurs in the nonlinear system in the presence of certain amount of input signal and added noises. For a nonlinear system, adding certain amount of noises can enhance the output signal-to-noise ratio (SNR) for a small input signal entering in the system and thus boost the signal detecting capabilities. For example, the driven double-well system (discussed in the previous section 7.1 for dynamical HC) with added noise is a typical one of cases, which can exhibit SR. The presence of noise can produce resonance-like effect in two-state (or double-well potential) nonlinear system and therefore make small signal easier to detect. Hence SR is apparently counterintuitive in nature, because in the usual sense the presence of noise is unwanted as it deteriorates the SNR capability. The phenomenon of SR exhibits deterministic kind of behavior in the presence of randomness which is in contrast to the phenomenon of chaos, where a randomness-like behavior is observed for deterministic systems. The phenomenon of SR is quite robust and its existence has been found in several areas of physical and life/medical/natural sciences, e.g., optical systems, electronic and magnetic systems, biological/neuronal systems, and climatic cycles [173,174]. The phenomenon of SR is also exhibited in simple electronic circuits (e.g., Schmidt trigger circuit), laser systems (e.g., semiconductor diode laser and bidirectional ring dye laser), thermally induced optical bistability in semiconductor materials [174-177], and neurophysiological systems [178] etc. In the literature a few reviews are available on the topic of SR [173,174].

The underlying mechanism behind SR could be understood (following Ref.[173]) with a very simplified model consisting of an over damped particle of mass m (due to a viscous frictional force and the damping constant related to the frictional force is characterized by the parameter f), making motion in a symmetric double potential well U(x), as shown in Fig.7.4. The particle is also coupled to a heat sink. Hence the fluctuational forces act on the particle due to such coupling to the heat sink. This model is quite popular in the study of reaction-rate theory [173,179]. The presence of fluctuational forces are responsible for inducing transitions between the adjacent potential wells. The rate of such transition is given by the Kramers rate formula [180]: $r_{KM} = \frac{\omega_m \omega_b}{2\pi f} \exp(-\Delta U/D_n)$. In this formula, $\omega_m = \sqrt{U''(x_m)/m}$ corresponds to angular frequency of potential at its minima at $\pm x_m$ and $\omega_b = \sqrt{U''(x_b)/m}$ is the angular frequency at the top of the barrier situated at x_b; ΔU represents the height of the potential barrier which separates the two minima of potential wells. The parameter D_n is the measure of noise strength such that $D_n = k_B T$, in which k_B is Boltzmann constant and T is the temperature. Application of a weak periodic force on the particle causes movement of the double-well potential to go up and down in an asymmetrical manner and thus inducing an increase and decrease of the potential barrier height in a periodic manner. The applied periodic force is not strong enough to cause the particle to jump from one well to another. However, the process of noise-induced hopping between the wells and the applied weak periodic force can synchronize with each other, provided the condition of $1/r_{KM} = T_\omega/2$ is satisfied, where T_ω is the period of the oscillatory (or periodic) weak force. Hence, the phenomenon of SR in a symmetric double-well potential is perceived by synchronized hopping events between the potential minima due to certain added noise with a weak periodic force [181]. For a selected T_ω, the synchronization can be achieved by changing the noise level D_n^{max} in accordance with the condition $1/r_{KM} = T_\omega/2$.

In a recent experiment, SR was demonstrated [48] in the AOB system with three-level atomic medium in Λ-type configuration of its levels (as shown in Fig.1.9(b)), contained in an optical ring cavity. This composite system is an ideal double-well potential system with controllability, as described in Chapter 3. This AOB system was driven by a weak periodic signal beam and an additive Gaussian white noise source with variable amplitude. This

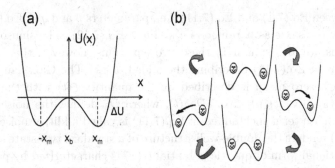

Fig. 7.4 (a) Sketch of the double-well potential $U(x) = (1/4)bx^4 - (1/2)ax^2$. The minima are located at $\pm x_m$, where $x_m = (a/b)^{1/2}$. The two minima are separated by a potential barrier with the height given by $\Delta U = a^2/(4b)$. The top of the barrier is located at $x_b = 0$. (b) In the presence of a periodic driving, the double-well potential $U(x,t) = U(x) - A_s x \cos(\Omega t)$ is tilted back and forth, thereby raising and lowering successively the potential barriers for the right and the left wells, respectively, in an antisymmetric manner. A suitable amount of noise, e.g., when the period of the driving force approximately equals to twice the noise-induced escape time, allows synchronized hopping to the globally stable (lower) state (adopted from [173] with permission).

system has exhibited a much better output SNR, showing a typical characteristic signature of SR. A good qualitative agreement between the experimentally measured results and the theoretical findings using the generic model was obtained [48], as will be described in the following.

7.2.1 *General theoretical model*

The equation describing the generic model of SR can be written as [5,48,173,174]

$$\dot{x}(t) = -U'(x) + A_s \cos(\Omega_s t + \phi_s) + \xi(t), \qquad (7.11)$$

in which U(x) is the reflection symmetric (double-well) potential given by the expression

$$U(x) = -\frac{a}{2}x^2 + \frac{b}{4}x^4, \qquad (7.12)$$

where A_s, Ω_s, and ϕ_s are the quantities defining the signal amplitude, signal frequency, and a simple phase factor, respectively. In Eq.(7.11), the dot (.) on top of the variable x represents the time derivative and apostrophe mark (') represents the space derivative. This equation (without the last noise term) was also used in describing the phenomenon of dynamical HC in AOB system as discussed above in subsection 7.1.1. Note the similarities between Eq.(7.1) and Eq.(7.11), i.e., parameters p and q in Eq.(7.1) are exactly the same as parameters a and b in Eq.(7.12). The sinusoidal driving terms are also similar in these two equations, however, there is an extra terms in Eq.(7.11) describing the added noise. The Gaussian white noise with a zero mean is described by the quantity $\xi(t)$ with the correlation function $< \xi(t)\xi(0) >= 2D_n\delta(t)$, where D_n defines the noise amplitude. The noise term such added in Eq.(7.11) is called additive noise. Note that $U(x)$ renders the double-well structure of a standard two-state system. The functional form of quartic potential $U(x)$ is characterized by parameters a and b, or to be more specific, the potential $U(x)$ critically depends on the parameters a and b. In the absence of noise term, Eq.(7.11) can be interpreted as an equation of motion for a driven anharmonic oscillator which is overdamped, having third-order Kerr nonlinearity [48]. When a periodic signal is applied to the system along with an added noise source, the SNR of the system output shows a maximum at a certain noise strength D_n, and it decreases on either side of this special value of D_n. So, the SNR in the output when studied as a function of D_n has a resonance-like feature, which clearly brings out the basic characteristic features of the SR phenomenon [5,48,173,174]. It is possible to have the output SNR exceeding the input SNR under specific conditions of parameters, which could provide very fruitful applications of SR in certain practical situations.

The generic Eq.(7.11) can be solved for arbitrary parametric conditions by numerical integration, so that extensive studies of SR phenomenon in such general double-well potential system could be accomplished. The phenomenon of SR is quite rugged because it does not depend on detail physical configuration of the system. For example, the basic SR feature is independent of the exact form of the potential used in Eq.(7.11) [48], which allows one to understand the prime characteristics of SR, at least qualitatively, by this simple physical model.

The input field associated in a normal AOB system changes in an adiabatic manner (as shown in Chapter 3), so the description of such system in terms of rate equations is justified [173,174]. Such rate equations can describe the dynamics of hopping events between the two states of the double-well system (located at x_+ and x_-). When the periodic signal input is applied to this system, it causes change to the depths of the two potential wells in an asymmetric manner (Fig.7.4). Under this condition the noise present in the system initiates transitions to occur randomly from one minimum to another in the double-well potential. In this situation the rate equations for probabilities p_+ and p_- of finding the system around minima x_+ and x_- are given by [5,48,173,174]

$$\dot{p}_+(t) = -S^+ p_+ + S^- p_-,$$
$$\dot{p}_-(t) = -S^- p_- + S^+ p_+. \tag{7.13}$$

Using Kramers equation [170], one can find the hopping rate S^+ (S^-) from the state x_+ to state x_- (state x_- to state x_+). According to the Kramers equation such rates are given by $\sim \exp[-2W^{u,d}/D]$, in which the quantity $W^{u,d} = W^0(1 \pm a_0 \cos(\Omega t))$ defines the time-dependent potential barrier. W^0 specifies the barrier height, a_0 is the amplitude of applied signal, and D is related to the noise variance. The autocorrelation function and power spectrum, can be calculated with the help of Eq.(7.13) from which one can obtain the SNR, given by [5,48,173,174]

$$SNR = (J^2/D^2) \exp[-2W^0/D]. \tag{7.14}$$

Here J is related to the intensity of the signal, as well as the height and width of the potential barrier. The parameter J could be estimated using the expressions for barrier height $(= a^2/4b)$ and barrier width $(= 2\sqrt{(a/b)})$ [48,173,174]. The parameters a and b, defined in Eq.(7.12), can be calculated solving the density-matrix equations Eq.(1.58) along with Eq.(3.19), and approximately estimated by analyzing the experimentally observed AOB curves. The calculated SNR curves are shown in Fig.7.5(b) for different values of signal amplitude, which are qualitatively in good agreements (curves A, B, and C are for $J = 2.737$, 5.477, and 7.303, respectively) with the experimental results for the similar parameters displayed in Fig.7.5(a), will be discussed in the subsequent section [5,48].

7.2.2 *Experimental demonstration of stochastic resonance*

The experimental arrangement used for demonstrating SR phenomenon was very similar to the experimental setup used for demonstrating three-level AOB as discussed in Chapter 3 (see Fig.3.14). The only difference between the two arrangements was an additional Gaussian white noise source connected to the EOM in the case of SR experiments. Note that the purpose of the EOM is to modulate and control the intensity of the input beam. The three-level atomic system in Λ-type configuration as shown in Fig.1.9(b) was realized by using the energy levels of the D_1 line of ^{87}Rb atom. The experiment began with acquisition of suitable AOB hysteresis curve. The

selected experimental parameters produced a refractive AOB curve. Note
that for the dispersive (refractive) AOB hysteresis curve of a two-level system, the parameters 'a' and 'b' in Eq.(7.12) represent the difference between
the frequencies of cavity resonance and the probe laser field, and the third-
order nonlinear susceptibility of the medium, respectively [169]. However,
in the case of this three-level AOB system, parameters 'a' and 'b' are closely
dependent on the probe laser frequency detuning Δ_P, the coupling laser frequency detuning Δ_C and the coupling laser power P_C etc (see Eq.(7.10) and
discussion above it in subsection 7.1.1) [48]. The input probe laser power
going inside the optical cavity was then adjusted to a value so that it fell
approximately in the middle of the observed hysteresis curve for AOB. After that a weak sinusoidal voltage signal of frequency at 150 Hz along with
the Gaussian broadband noise generated by an arbitrary waveform generator were added to the EOM, which provided the additive noise as described

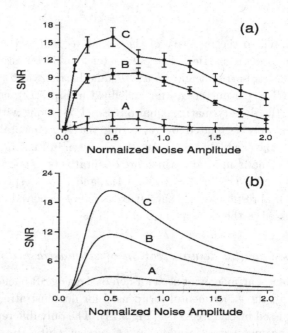

Fig. 7.5   (a) Experimentally measured output SNR as a function of input noise amplitude. The frequency of the sinusoidal signal is 150 Hz. Here, curves A, B, and C are for
three different values of signal amplitudes corresponding to 10 μW, 30 μW, and 45 μW,
laser power variation about the point of operation of the cavity input laser, respectively.
The x-axis represents twice the standard deviation of noise amplitude divided by the
width of bistable region. (b) The theoretical predictions from the generic model of SR
(Eq.(7.11)) with similar parameters (reprinted from [48] with permission).

by Eq.(7.11). Thus the weak signal and noise rode on the fixed operating input power level. The input and output optical signals (i.e., light beams going in and coming out of the cavity) were monitored using APD detectors. The waveforms of the input and output signals from the optical cavity were obtained in a time duration of 1 sec, which were digitally stored as a time series. Using fast-Fourier transform on the digitized data, the power spectrum of the optical signals could be obtained. This process was repeated many times so that the averaged power spectrum could be attained [48]. In order to quantify the SR phenomenon, the definition used for the output SNR was the ratio of the magnitudes of the power spectrum at the signal frequency, and the noise level in the absence of the input signal. The SNR so defined is called the classical narrowband SNR in literature [173]. There is yet another definition of SNR called the wideband SNR, which is calculated by finding the ratio of the total power in the deterministic part of the signal and its harmonics in the power spectrum to the total power in the noise part [48]. The limitation to use a broad bandwidth noise source in such experiment was set by the frequency response of the EOM.

The experimentally measured output SNR curves as a function of the noise strength for the three-level AOB system with different signal amplitudes are displayed in Fig.7.5(a). In this figure, the y-axis accounts for the SNR in linear scale and the x-axis represents normalized noise strength. The normalized noise strength is defined as the ratio of twice the standard deviation of the noise strength to the width of the observed AOB hysteresis cycle. The reason for normalizing the x-axis is just for the sake of convenience because the maximal SNR is usually found to be at the noise variance equal to half the barrier height (approximately) of the double-well potential, i.e., at half the width of the AOB hysteresis cycle [48]. In the experiment it was quite easy to control the width of AOB hysteresis cycle and thus allowed the observation of SR phenomenon for several different AOB hysteresis curves. The experimental parameters used in Fig.7.5 were T= 68 ^{0}C, P_C = 12 mW, Δ_C = 150 MHz, Δ_P = 100 MHz, and cavity detuning Δ_θ = 50 MHz. The curves A, B, and C are plotted for three different signal amplitudes as described in the caption, with other parameters the same. A common feature in all three curves can be seen easily, i.e., the SNR first goes up with an increasing noise strength, and after reaching a maximum, it reduces down as the noise strength is increased further. All the measured curves exhibit such resonance-like behavior. This kind of resonance pattern was verified for several different experimental conditions of observed AOB hysteresis curves [5,48].

A generic double-well potential system as discussed above in subsection 7.2.1 is a quite suitable model to describe the experiment and simplifies the understanding of the fundamental mechanism for SR phenomenon. This is because the Fokker-Planck equation used to model the overdamped motion of a particle in a double-well potential is very appropriate to analyze the phenomenon of AOB [1]. The AOB hysteresis curve, which has stable upper and lower branches, can be mapped to the states of two wells in the double-well potential. By applying a periodic signal in the AOB system at cavity input, the depths of these two wells change periodically, and then the added external noise works in collaboration with the signal to induce the particle jumping from one well to another in the double-well potential. For AOB, this means switching between two stable states of the hysteresis curve. When noise strength is small, the hopping rate between wells is quite low. By increasing the noise strength, the hopping events are enhanced and the SNR acquires a maximum for a certain value of noise strength with respect to the amplitude of the applied signal. If the noise strength becomes too large, then the rate of hopping also gets quite large, but the signal is lost in the large background noise, therefore reducing the SNR [48]. The simple model (presented in Eq.(7.11)) agrees well with this explanation as shown in Fig.7.5(b) corresponding to parameters used in Fig.7.5(a).

The SR phenomenon was demonstrated for a wide range of experimental parameters in the three-level AOB system, leading to the conclusion about the generality and robustness of this interesting phenomenon for a great range of physical situations [48]. The gain in SNR (G), as a function of noise strength (Fig.(7.6)) also shows a similar trend as the output SNR in Fig.(7.5), i.e., there is a peak in the gain G at certain noise strength (a resonance-like behavior). Here $G > 1$ is interpreted as an evidence of strong cooperative phenomenon in the system such that incoherent noise power is contributing to enhance the coherent output signal. The presence of EIT medium inside the cavity generates coherence in the AOB system that may further enhances the above cooperative phenomenon.

The adiabatic approximation for the AOB system under consideration is justified under the condition of $\Omega << a$, where 'a' is as defined in Eq.(7.12). This fact originates from the application criterion of Kramers rate formula, which requires the probability density within a well to be approximately at

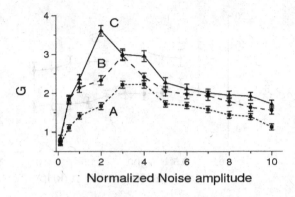

Fig. 7.6 Experimentally measured SNR gain G as a function of input noise amplitude. The curves A (dotted), B (dashed), and C (solid) are for the three signal amplitudes same as in Fig.(7.5) (reprinted from [48] with permission).

equilibrium [1,48]. If the signal frequency is slower than the rate for equilibration of probability within a well, the modified Kramers rate could be properly realized. For low values of Ω, the adiabatic condition is well satisfied for the dispersive AOB settings in the experiments. The dependence of the output SNR on the frequency of applied weak signal was measured for fixed amplitudes of the signal and noise going into the optical ring cavity. The signal frequency was changed from 0.15 kHz to 2.0 kHz. The remaining experimental parameters were set as T $= 74$ ^{0}C, $P_C = 12$ mW, $\Delta_C = 100$ MHz, $\Delta_P = 40$ MHz, and $\Delta_\theta = 50$ MHz. From Fig.7.7 it is clear that the output SNR decreases as the signal frequency increases within the reasonable experimental uncertainties. In this figure curves A and B represent two different signal amplitudes as described in the figure caption. The observed results also ratify the prediction of the generic model for SR (Eq.(7.11)) discussed above, i.e., for fixed noise and signal amplitudes there is a slight decrease in the output SNR with the increase in signal frequency. This is because as Ω crosses the adiabaticity condition limit, the applicability of Kramers rate formula is no longer valid [173,48]. For experimental investigation of the SR phenomenon, the passive three-level AOB system has provided an ideal platform to have a quartic potential well (two-state) system, since the shape and thresholds of the hysteresis curve can be easily manipulated and controlled by various experimental parameters as shown and discussed in Chapter 3. Hence the double-well potential in terms of its height and separation of the two minima can be modified easily by controlling laser beam parameters, which can be used to further explore several

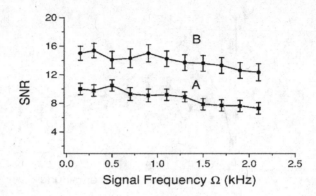

Fig. 7.7 . Experimentally measured output SNR as a function of signal frequency for fixed normalized noise amplitude and signal amplitude. Curves A and B are for two different signal amplitudes corresponding to 30 and 50 μW laser power variation about the point of operation of the cavity input laser, respectively (adopted from [48] with permission).

interesting and unknown qualities of SR phenomenon. Since it is comparatively much simpler to manipulate the linear absorption/dispersion and Kerr nonlinear properties of such three-level AOB system, one can systematically explore the effects of noise transfer in this nonlinear AOB system, as well as noise-assisted flipping between the bistable states [48]. The experiments discussed above have used additive noise, but the SR phenomenon could also be observed in the presence of multiplicative random noise. In a recent experiment such multiplicative random noise was applied onto the cavity frequency detuning (by driving the PZT on which a cavity mirror was mounted with a random noise generator) with a modulated signal on the cavity input light, and the results showed many interesting features [182].

7.3   Noise-induced Switching between Bistable States in a Three-level Atomic Optical Bistability System

The AOB systems can be well described under the deterministic conditions [1,2,4,5] as discussed in the previous Chapters 3-6. On the other hand, the responses of these systems to fluctuations in the experimental parameters, such as the cavity field intensity or laser frequencies are still challenging to explore. In the presence of noises, the dynamical switching between the two bistable states of such AOB systems can be greatly influenced.

In the previous section 7.2, we have described the phenomenon of SR, where externally added noise in a nonlinear system (e.g., a three-level AOB system) can enhance the SNR by synchronizing the signal and noise in the system [48]. Switching between bistable states induced by intrinsic noises can also occur when the AOB system is below its deterministic threshold value. Several significant works of this type on lasers [1,2,183], hybrid electro-optical systems [184], passive all-optical double-cavity membrane system [185], can be found in the literature.

7.3.1 *Basic theory of noise-induced switching*

There have been different ways to describe noise-induced transitions in non-linear optical systems [173,174]. In the following a simplified theory of noise-induced switching between the two bistable states in the three-level (AOB) system will be outlined. The AOB system consists of an optical cavity containing vapor of ^{87}Rb atoms in Λ-type configuration (Fig.1.9(b)), which forms EIT medium with controllable absorption, dispersion and nonlinear properties, as shown in Chapter 1. As discussed previously, this medium possesses greatly enhanced Kerr-type nonlinearity due to the light-induced atomic coherence among the energy levels in the system [4]. If there are fluctuations in the intensity and/or frequency of the intracavity field or in the length of the optical cavity, it will bring additional changes in the magnitudes of induced atomic coherences in such three-level EIT medium. It is clear through the discussions in Chapter 1 that the change in the induced atomic coherence can change the index of refraction of the medium, which eventually leads to modification in the round-trip cavity phase and therefore initiates the switching in AOB processes between the bistable states. The noise-induced switching is a physically different phenomenon from the dynamic instability [46,142] discussed in Chapter 5. The dynamic instability in AOB systems is periodic in nature and can be explained by deterministic formulation of the problem [1,2].

The underlying physical mechanism for noise-induced optical switching can be explained with the help of a simple model of coupled atom-cavity system. The composite system of atom-optical ring cavity exhibits dispersive AOB, because the cavity is filled with a nonlinear atomic vapor (a Kerr-medium) whose refractive index is directly proportional to the cavity field intensity. In this situation the quantity $S(\omega_P) \equiv I_{out}/I_{in}$, called cavity response function (derived in Chapter 1, Eq.(1.48)) of the optical ring cavity of length l containing an atomic vapor inside a glass cell of length L, is given by the following expression [49,62]

$$S(\omega_P) = \frac{T^2\kappa}{(1 - R\kappa)^2 + 4R\kappa \sin^2(\theta/2)}, \qquad (7.15)$$

in which the reflection and transmission coefficients of the input and output cavity mirrors are given by R and T, respectively, such that $R = 1 - T$. The Finesse of the empty cavity is given by $F = \frac{\pi\sqrt{R}}{1-R}$ (see Eq.(1.41)). The absorption caused by the medium is given by $\kappa \equiv \exp(-\alpha L)$, in which α is the absorption constant. The cavity field acquires a phase shift θ upon completion of a round trip through the cavity,

$$\theta = \frac{\omega_P l}{c} + (n_0 - 1)L\frac{\omega_P}{c} + L\frac{\omega_P}{c}n_2 I_P, \qquad (7.16)$$

where ω_P is the probe field frequency, I_P is the intensity of intracavity field, n_0 and n_2 are the linear refractive index and nonlinear Kerr-type index of the medium, respectively.

Since the hysteresis curve of dispersive (refractive) AOB is controlled by the nonlinear refractive index of the intracavity medium, the exact knowledge regarding time evolution of the intracavity round-trip phase is important in this case. The relaxation of the intracavity round-trip phase (which is induced by the nonlinear refractive index) is governed by the following equation (better known as Debye equation in the literature) [2,186]

$$\dot{\theta}_{NL} + (1/\tau)\theta_{NL} = \frac{X\omega_P I_P}{(1 - R\kappa)^2 + 4R\kappa \sin^2([\theta_{NL} + \omega_P t_R]/2)}, \qquad (7.17)$$

where θ_{NL} is the nonlinear phase shift accrued by the cavity field and given by $(\frac{\omega_P L}{c})n_2 I_P$ and $X = 2\pi n_0 l\beta T/c$ where β is some fixed parameter [186]. The above equation is valid in the small-cavity limit, i.e., when the cavity round-trip time t_R is small in comparison to the relaxation time (τ) of the medium and there are no field fluctuations of any kind. The validity of the small-cavity limit is justified for the EIT medium contained in an optical cavity during the experiment [49]. Inclusions of the cavity field frequency and intensity fluctuations $(\delta\omega_P$ and $\delta I_P)$ about some mean values $(\omega_P$ and $I_P)$ leads to the modification of Eq.(7.17) as

$$\dot{\theta}_{NL} + (1/\tau)\theta_{NL} = \frac{X(\omega_P + \delta\omega_P)(I_P + \delta I_P)}{(1 - R\kappa)^2 + 4R\kappa \sin^2([\theta_{NL} + (\omega_P + \delta\omega_P)t_R]/2)}.$$
$$(7.18)$$

We define $(\omega_P + \delta\omega_p)t_R = -D + \omega_m t_R + \delta\omega_P t_R$, where $D = (\omega_m - \omega_P)t_R$ and ω_m is the cavity resonance frequency. The cavity resonance frequencies are given by $\omega_m t_R = 2\pi n_i$ (n_i : an integer) because of the cavity mode definition. Hence Eq.(7.18) can be rewritten as [49,186]

$$\dot{\theta}_{NL} + \theta_{NL} = \frac{Q(1 + \delta\omega_P/\omega_P)(1 + \eta)}{(1 - R\kappa)^2 + 4R\kappa \sin^2([\theta_{NL} + \delta\omega_P t_R - D]/2)}. \tag{7.19}$$

While writing Eq.(7.19), the time variable has been redefined by inclusion of τ in it (in other words normalizing it with τ). The other variables are defined as $Q = I_P/I_S$, $I_S = c/(2\pi n_0 l \omega_P \beta T \tau)$, where I_P is the mean value of the intracavity field intensity, $\eta(t) = \delta I_P/I_P$ and $\delta\omega_P/\omega_P$ are stochastic variables with zero mean value corresponding to the intensity and frequency fluctuations of the intracavity field, respectively [49,186]. In order to simplify Eq.(7.19) further, a new phase variable $\phi = \theta_{NL} - \theta_0 + \delta\omega_P t_R$ is introduced in the equation, where a steady-state solution of Eq.(7.19) (ignoring any fluctuations) is used to determine θ_0:

$$\theta_0 = K(Q, \theta_0). \tag{7.20}$$

The quantity $K(Q, \theta_{NL})$, appearing on the right hand side of Eq.(7.19) in the absence of any fluctuations, is characterized by the expression

$$K(Q, \theta_{NL}) = \frac{Q}{(1 - R\kappa)^2 + 4R\kappa \sin^2([\theta_{NL} - D]/2)}. \tag{7.21}$$

The evolution dynamics of this new phase variable ϕ (defined above) along with fluctuations is given by the following equation [49,186]

$$\frac{\partial \phi}{\partial t} = M_0(\phi) + M_1. \tag{7.22}$$

In Eq.(7.22) above, $M_0(\phi) = -\phi/W_0 + (\phi^2/2)K^{(2)}$, $M_1 = (\theta_0 + \omega_P t_R)(\xi + \dot{\xi}) + \theta_0\eta$, and $W_0 = 1/(1 - K^{(1)})$. The first and second derivatives of K with respect to θ_{NL} (at $\theta_{NL} = \theta_0$) are given by $K^{(1)}$ and $K^{(2)}$, respectively, and the frequency fluctuations of the cavity field are determined by the parameter $\xi = \delta\omega_P/\omega_P$. When the value of $K^{(1)}$ approaches to 1, the system is near its turning point and its response time becomes extremely long. This is because of the critical slowing down effect [1]. Both of these processes, i.e, the rapid noise fluctuations and critical slowing down, are present in Eq.(7.22), which allow one to study the combined effect in the switching process [186]. The analysis of the expression for M_1 clearly indicates that the terms depending on the frequency fluctuations are multiplied by a factor $\theta_0 + \omega_P t_R$, and, on the other hand, the term depending on the amplitude fluctuation is simply multiplied by a factor of θ_0 only. For a 37 cm long optical ring cavity, sustaining modes of optical frequencies, it is straightforward to estimate that $\omega_P t_R >> \theta_0$ by about a factor of $\sim 10^5$ [49]. So, it is easy to conclude that the frequency fluctuations in the laser beam and cavity length contribute much more over the laser amplitude fluctuation in the experiments. Another important thing to be noticed here is that the correlation time scale of fluctuations is much shorter than the deterministic time required for the relaxation of nonlinear phase shift. Equation (7.22) is the Langevin equation of the system under consideration and the corresponding Fokker-Planck equation [1,49,186] is given by

$$\frac{\partial P}{\partial t} = \frac{\partial}{\partial \phi}[\frac{\partial U}{\partial \phi}P] + D_F\frac{\partial^2 P}{\partial \phi^2}, \tag{7.23}$$

where P is the probability distribution of the process [1], and function $U(\phi) = \phi^2/(2W_0) - (\phi^3/6)K^{(2)}$.

Physically, Eq.(7.23) represents a trapped particle in a potential well $U(\phi)$ diffusing under the action of stochastic process with a diffusion constant D_F (details about D_F are given in Ref.[186]). In other words it governs the phase diffusion process and the minimum and maximum of the potential function $U(\phi)$ are situated at the positions $\phi = 0$ and $\phi = 2/(K^{(2)}W_0)$, respectively. Our original problem is the study of noise-induced transitions in an AOB system. Thus comparison or analogy with the original problem can be made by considering the solution of the deterministic steady-state Eq.(7.20) in the neighborhood of up-switching point of the AOB hysteresis curve. In Fig.7.8, a typical dispersive AOB hysteresis curve ABCD is plotted for the variables θ and Q, generated by Eq.(7.20). In the AOB hysteresis curve, branches AB and CD are the stable ones of the hysteresis curve but branch BC is unstable (having a negative slope). The requisite conditions for up-switching or upper threshold can be realized by solving the equations $\theta = K(Q, \theta)$ and $\frac{\partial \theta}{\partial Q} = \infty$, where $K(Q, \theta)$ is given in Eq.(7.21). Note that the subscript on θ has been omitted for the sake of convenience. The upper threshold or up-switching value (Q_s, θ_s) so obtained is designated as point B on the hysteresis curve displayed in Fig.7.8. In a quite similar manner the condition for down-switching or lower threshold of the hysteresis curve can be defined, which has been designated as point C on the hysteresis curve displayed in Fig.7.8. The difference between the upper threshold value and lower threshold value determines the width of the AOB hysteresis curve in terms of the distance between points C and B along Q-axis of Fig.7.8. Equation (7.20) can be solved in the close proximity of the switching points by making use of Taylor series expansion of $K(Q, \theta)$ to get

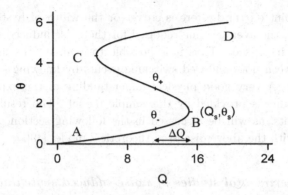

Fig. 7.8   Plot of θ (dimensionless) as a function of Q (arbitrary units) using Eq.(7.20) for $D/\aleph = 4.5$ ($\aleph = (1 - \kappa R)/(2\sqrt{R\kappa})$) showing a hysteresis cycle ABCD having stable branches AB and CD, and unstable branch BC. $\theta_\mp$ and ΔQ are defined in the text (reprinted from [49] with permission).

$\theta_\mp = \theta_s \mp \sqrt{(2\theta_s \Delta Q / K^{(2)})}$. In this expression of $\theta_\mp$, negative (positive) sign on the right hand side is for $\theta_-(\theta_+)$ and the quantity $\Delta Q = |Q_s - Q|$ in Fig.(7.8) measures the deviation of Q value from the upper threshold value Q_s [49]. The separation between the unstable and stable branches of the AOB hysteresis cycle in Fig.7.8 can be estimated in terms of the distance between the maximum and minimum of the potential function, given by $\theta_+ - \theta_- = 2\sqrt{(2\theta_s \Delta Q / K^{(2)})}$. With the push of a large noise spike, the particle resting at the minimum of the potential well can jump to the maximum. At this stage the particle can have two options, i.e., from the position of maximum it can either go back to the original position or make an exit from this well. In the AOB system shown in Fig.7.8, the system is initially on a stable branch point with phase θ_0 (near θ_-). Due to noise surge, it moves to the unstable point with a phase θ_1 (near θ_+), which is located on the unstable branch of the hysteresis curve. Once here then the system may go back to the stable point θ_0 or jump to the another stable branch, i.e., the upper branch of the AOB hysteresis curve, which accomplishes the process of noise-induced up-switching. For a particle initially situated at θ_0, the average escape time can be estimated to be [49,186]

$$T^e \cong \frac{c^{(1)}}{\sqrt{\Delta Q}} \exp[c^{(2)}(\Delta Q)^{3/2}], \qquad (7.24)$$

where quantities $c^{(1)}$ and $c^{(2)}$ are dependent on the experimental parameters and have positive values. When ΔQ is small, noise spikes can induce considerably large switching events. As ΔQ is increased by changing the operating point on the hysteresis curve (or the width of hysteresis curve is increased), the average time required for the noise-induced switching is significantly increased. Thus it is possible to estimate the average time interval between noise-induced switching events by knowing such particle escape time. A very good physical understanding of the experimentally observed results is provided by this simple model. The results obtained in experiments, as will be discussed in the following section, are in good agreement with the above discussed theoretical model [49].

7.3.2   *Experimental studies of noise-induced switching*

Experimental demonstrations of noise-induced switching were carried out in the AOB system consisting of three-level atoms (rubidium atomic vapor) in Λ-type configuration in an optical ring cavity [49]. The experimental system was quite similar to the one carried out to demonstrate controllability of hysteresis cycle characteristics of AOB (discussed in Chapter 3) [42] and the phenomenon of SR (discussed in subsection 7.2.2) [48] in AOB. More specifically, the energy levels of the D_1 line of ^{87}Rb atom were employed to form the required three-level system in Λ-type configuration (see Fig.1.9(b)). The experimental setup was the same as the one depicted in Fig.3.14, which was used for demonstrating the controllability of AOB, except with a white-noise source applied to the EOM before the cavity input. In the experiment it was possible to observe a variety of AOB hysteresis curves by simply changing the intensities and frequency detunings of the control and probe laser beams, temperature of the Rb cell, and the cavity frequency detuning Δ_θ. The EOM was driven by a triangular ramp voltage source in order to modulate the intensity of cavity input field, which is used to observe the typical bistable curve. After obtaining the required dispersive (refractive) AOB hysteresis curve, the triangular ramp voltage on the EOM was switched off. The cavity input laser intensity level was then adjusted by the voltage on the EOM in such a way that it remained at a level almost in the middle of the observed AOB hysteresis curve. The average linewidths of the diode lasers were in the range of a few MHz in the experiment. These diode lasers were extended cavity type with their amplitude and frequency stabilized. However, there were still some residual phase/frequency and intensity noises associated with these lasers, responsible for inducing fluctuations in the nonlinear refractive index of the intracavity medium, thus causing the switching events between the bistable states of AOB. It was established by experimental measurements that the influence of intensity fluctuations in diode lasers to atomic systems in comparison to the frequency fluctuations is about 10-100 times smaller. This means that the mechanism responsible for the noise-induced switching is predominantly caused by the frequency fluctuations of diode lasers and not the intensity fluctuations. The possibility of frequency jitters in the optical ring cavity mode could be attributed to the linewidth of the third diode laser used for locking the optical cavity and the electronic noise in the feedback locking circuit. The frequency fluctuation of the third diode laser was of the same order of magnitude as that of the probe laser and it was estimated to be about few MHz. The electronic noise could be assumed to be small. Hence for the experiment the cavity frequency jitter was also an integral part of the frequency fluctuation in association with the frequency fluctuation of the circulating probe laser in the ring cavity. The output intensity from the cavity was measured using an APD detector and the signals were stored digitally in an oscilloscope. Such fast detector was sufficient to observe switching events caused by the fluctuating refractive index of the medium. In order to obtain an averaged switching response of the optical cavity output for given experimental conditions, each experimental measurement was repeated several times [49] in a given time interval. An experimentally observed switching time-trace of the cavity output is shown in Fig.7.9(a). The experimental parameters used to obtain that time-trace are as follows: atomic sample cell temperature T= 68 ^{0}C, coupling (control) laser power $P_C = 12$ mW, probe laser frequency detuning $\Delta_P = 50$ MHz, and cavity detuning $\Delta_\theta = 50$ MHz. Plots (a), (b), and (c) in Fig.(7.9) represent three different AOB hysteresis curves with decreased width and reduced coupling (control) laser frequency detunings of $\Delta_C = 55$, 35, and 25 MHz, respectively, but keeping other experimental parameters the same. It is clear from these curves that by altering the operating point of the input intensity on the hysteresis curve towards the upper threshold (or up-switching point) of the AOB, there is an enhancement in the occurrence of switching events to the upper metastable state of the hysteresis curve in the same time interval. Note that the operating point is effectively moved nearer to the up-switching point by reducing the width of the AOB hysteresis curve, but not too close to the up-switching point so that the critical slowing down takes over and the system goes sluggish. Due to the impulse provided by the noise peak a sharp spike nearly up to the upper state is observed, which quickly decays down to the lower state following the asymmetric potential $U(\phi)$ curve and the discussion given above [49]. Also, negative noise spikes pull down system to lower state.

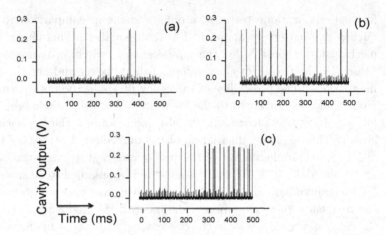

Fig. 7.9   Typical noise-induced switching traces observed in cavity output intensity as a function of time in the AOB system with experimental parameters mentioned in the text. Curves (a), (b), and (c) are for three different coupling laser detunings, respectively, with all other parameters fixed (adopted from [49] with permission).

Another interesting physical mechanism also contribute to the switching phenomenon in such optical cavity containing a nonlinear medium. The cavity linewidth narrowing takes place due to the presence of intracavity EIT medium as discussed in subsection 3.1.1. When the refractive index changes by a small magnitude in the cavity, the optical cavity mode moves by an amount in the order of its linewidth leading to a switching from no cavity transmission or zero intensity state (system in the lower branch of AOB hysteresis curve) to cavity transmission or non-zero intensity state (system in the upper branch of AOB hysteresis curve). At the same time, the frequency stabilization factor of the cavity mode, given by the expression $\frac{\omega_P L}{l} \frac{\partial (Re(n))}{\partial \omega_P}$, where n is the total refractive index, acts on the system and get it back to its initial condition [5,49], so that the effect as displayed in Fig.7.9 could be observed. To further quantify the noise-induced switching phenomenon more experiments were performed [49]. In one such experiment the occurrences of the cavity output intensity switching to the upper branch of the AOB hysteresis curve as a function of AOB width were measured, which are displayed in Fig.7.10. The experimental parameters were $T = 68\ ^0C$, $P_C = 12$ mW, $\Delta_P = 50$ MHz, and cavity detuning $\Delta_\theta = 50$ MHz. The width of AOB hysteresis cycle was controlled by making changes in Δ_C value from 25 MHz to 65 MHz. The operating point of

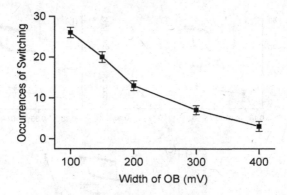

Fig. 7.10 Occurrences of switching events in cavity output intensity as a function of width of the AOB hysteresis curve. The experimental parameters: T = 68 ^{0}C, P_C = 12 mW, Δ_P = 50 MHz, and cavity detuning Δ_θ = 50 MHz. AOB widths were controlled by using Δ_C in the range from 25 MHz to 65 MHz (reprinted from [49] with permission).

the cavity input power level was set always at the midpoint of each AOB hysteresis curve. As the width of the hysteresis curve increases (Fig.7.10), the number of occurrences of the switching events goes down. This is because the spikes produced by the noise are less likely to be big enough in comparison to the increase in the half width of AOB curve, and hence switching can not take place. The average time required for the switching events to occur increases with the widening of the hysteresis loop, which is in good agreement with the theoretical prediction discussed above in subsection 7.3.1 [49,5].

The noise-induced switching phenomenon was further investigated quantitatively by using three AOB hysteresis curves of different widths as shown in Fig.7.11. The parametric conditions used in the experiment were T = 70 ^{0}C, P_C = 10 mW, Δ_P = 60 MHz, and cavity detuning Δ_θ = 40 MHz. The plots (a), (b), and (c) in this figure are for Δ_C = 38, 52, and 67 MHz, respectively. The parameter on the x-axis is X, which is defined as the ratio of the operating point with respect to the lower threshold of the AOB hysteresis curve to the total width of hysteresis cycle in the AOB curve. The main idea behind these measurements is to allow the system to sit on different operating points in each of the hysteresis curves and then observe the switching events. These hysteresis curves are shown in the insets of Fig.(7.11). The observed noise-induced switching occurrences are plotted with respect to the relative positions of operating point within

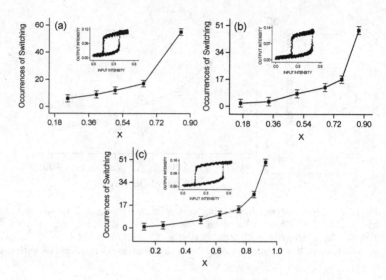

Fig. 7.11   Occurrences of switching events as a function of X for three different hysteresis curves of AOB. The experimental parameters are: T= 70 ^{0}C, $P_C = 10$ mW, $\Delta_P = 60$ MHz, and cavity detuning $\Delta_\theta = 40$ MHz. Curves (a), (b), and (c) are for $\Delta_C = 38, 52,$ and 67 MHz, respectively. Here X is defined as the ratio of the operating point (with respect to the lower threshold) on the optical bistability curve to the width of AOB hysteresis curve (reprinted from [49] with permission).

the AOB curve in Fig.7.11. For the hysteresis curve with a small width (Fig.7.11(a)), there is a large number of switching events at low operating points (far from threshold) in comparison to the hysteresis curves shown in Figs.7.11(b) and 7.11(c) having broader hysteresis widths. The strengths of the large noise spikes are good enough to make the cavity field jump from the lower state to the upper state of the AOB hysteresis curve with a narrower width of the bistable region. However, when the operating points reach near the upper threshold point (but not too near so that critical slowing down takes over), switching events increase substantially in all the AOB hysteresis curves and thus show similar behavior.

In the experiments discussed above, noise-induced switching in the passive AOB system of three-level rubidium atoms in Λ-type configuration inside an optical cavity was considered. The intracavity EIT medium was driven by a cavity field which had some intrinsic fluctuations in its frequency as well as intensity. A control field also interacted with the intracavity medium. The fluctuations of cavity field transfer to the generated atomic coherence in the medium, which enforce enhanced nonlinearity to fluctuate, initiating the phenomenon of random switching to occur between bistable states of AOB. There could be a potential application of such phenomenon in realizing pseudo random number generation useful for quantum communication and cryptology. Further experimental and theoretical investigations should be done to better understand the roles of various types of the noises on the switching events in such AOB systems. The noise-induced switching discussed here is based on the classical noises associated with laser sources in terms of their frequency and amplitude fluctuations. It could be possible to observe such switching phenomenon with quantum fluctuations using similar experimental system, which will be of great importance to establish the possibility of investigating quantum tunneling phenomenon in the double-well potential systems.

Chapter 8

Conclusion and Outlooks

The AOB and related effects in two-level systems have been extensively studied in eighties and many interesting theoretical results were predicted [1,2]. For instance, the mean field model of AOB was discovered and successfully applied in several situations including mixed absorptive and dispersive AOB using semiclassical and quantum statistical mechanics approaches. The mean field model [86] was also used in describing the switching characteristics in two-level AOB systems. Another success story of mean field model using quantum statistical formulation was in calculating spectra of transmitted and fluorescent light beams [80,187] and predicted quantum effects including photon statistics of the transmitted light [1] and squeezed state [188]. Also, several experimental demonstrations of AOB and related effects were reported by researchers in different groups using various experimental systems. Several ingenious ideas were applied for circumventing the Doppler effect to facilitate the observation of AOB in two-level atomic systems, such as using atomic beams or buffer gases in atomic vapor. These experimental schemes were then applied to observe instabilities in AOB systems. Although chaos and stochastic resonance behaviors were theoretically predicted in two-level AOB systems, they have not been experimentally observed in such systems, mainly due to the difficulty of reaching the desired parameters. However, the main difficulty in AOB experiments using two-level systems was the lack of good experimental controls in all those experiments, i.e., it was very difficult to produce the desirable AOB hysteresis curves and get into certain parametric regions in those experiments.

The uses of three-level atomic systems as the intracavity media have opened a new era in studying AOB and related dynamical effects in terms of the controllability of AOB hysteresis cycles. The central idea of controllability in three-level EIT medium lies in the controllable optical properties, e.g., absorption, dispersion and nonlinearities due to the induced atomic coherence in such medium, which eventually control all the phenomena related to the three-level AOB. The three-level EIT system consisting of hot rubidium atomic vapor (contained in a glass cell) in Λ-type configuration was used as the intracavity medium to show the controllability of AOB hysteresis curve. This three-level AOB system has been facilitated to produce diverse patterns of hysteresis loops including shape changes (from absorptive to dispersive AOB, and tunable width and thresholds etc) and backward hysteresis cycles with the help of available control beam parameters. The atomic optical multistability was also produced by this system in a well-controlled manner using several experimental parameters. This composite AOB system with EIT intracavity medium and the optical ring cavity was then used in the experimental demonstrations of the controlled optical dynamic instability and chaos via period-doubling route. The experimental results obtained for showing optical instability and chaos were in excellent agreements with the theoretical calculations over a wide range of parameters using the model developed to govern the dynamics of such composite atom-cavity system. The experimental parameters used for controlling AOB, AOM, optical instability and chaos included the laser powers and frequency detunings of the cavity (probe) and coupling (control) fields, cavity detuning and the atomic density. It is such external controls of experimental parameters in this three-level AOB system that makes it superior over the previously used experimental AOB systems, where experimental flexibility was restricted since only one laser beam was used for experiments. Such three-level AOB systems with controllable steady-state curves have been used to demonstrate controlled all-optical switching between two bistable states, in which simply changing the frequency or intensity of the control laser beam, the intensity of the cavity field was turned 'on' and 'off' very efficiently. Such three-level AOB system made it quite easy to systematically study the phenomenon of dynamical hysteresis when driven by a high frequency periodic optical signal. The phenomenon of stochastic resonance was also studied using this system when it was simultaneously driven by a periodic signal with an added random Gaussian white noise, which has led to an enhanced output SNR or large gain in SNR, confirming the cooperative effects between signal and noises in this double-well potential system with enhanced nonlinearity. Another remarkable phenomenon of noise-induced switching in double-well three-level AOB system has been studied systematically, again due to the controllability of AOB in such a system. The physical mechanism associated with this phenomenon is the intrinsic frequency and intensity fluctuations in cavity field, which can induce fluctuations in atomic coherence and thus influence the enhanced nonlinearity, resulting in random switching between the bistable states of the AOB system. The noise-induced switching is a random process and hence could be used in realizing of pseudo random number generation, which is an essential ingredient in quantum communication and cryptography applications [189].

Many other nonlinear optical processes, such as four-wave mixing, six-wave mixing harmonic generations and Raman processes, etc are shown to be greatly enhanced by the presence of induced atomic coherence in multilevel EIT systems. In the past twenty years, there have been large amount of theoretical and experimental works done with multilevel, especially three-level, EIT systems to investigate the modified linear and nonlinear optical properties. When such coherently-prepared EIT media are placed inside optical cavities, many interesting and novel steady-state, as well as transient effects can appear, as discussed in this monograph. Such intracavity medium allows one to perform systematic studies of some fundamental phenomena and also provide the needed information for possible applications of multilevel EIT systems in all-optical switching, all-optical buffering, and generating optical solitons etc, which are very useful in all-optical computation and networking.

Another interesting direction of research in such three-level AOB systems with induced atomic coherence is to observe quantum tunneling. The AOB system, having a double-well potential, should be an ideal platform to investigate quantum tunneling phenomenon between the two stable states (wells). Such important and interesting effect has not been observed in the two-level AOB systems inside an optical cavity, since the predicted tunneling time would be extremely long making the experimental observation unrealistic. However, in the three-level AOB systems, the bistable curve, or the double-well potential, can be controlled easily by the experimental parameters, such as the control (coupling) beam frequency detuning and intensity, cavity detuning, probe beam frequency detuning and atomic density, as shown in Chapter 3. The enhanced nonlinearity and modified absorption and dispersion properties can generate AOB with very low light intensities. So, it is feasible to modify the barrier (both width and height) between the two wells and therefore significantly reduce the tunneling time, and thus it may be possible to observe the quantum tunneling phenomenon in the three-level AOB system.

Look forward for the novel experiments in quantum and nonlinear optics, there are several interesting and worth to explore areas which can be further investigated with multilevel atomic systems exhibiting the phenomenon of EIT, especially in optical cavities. One of such directions is to achieve efficient nonlinear optical processes with single photons. Some single-photon optical processes, for example single-photon switching [163] and single-photon quantum networking [190], have been theoretically studied in EIT enhanced systems. Correlated photon pairs and entangled photons [191,192] have been generated in multilevel EIT atomic systems. Using entangled or correlated photon pairs with engineered efficient nonlinear interactions at single-photon level could facilitate to realize logic gate operation [163], storage devices for photons [191], quantum networking and quantum information processing [190]. By placing such coherently-prepared atomic system inside an optical cavity, the rate of photon pair generation has been greatly increased [193] and bright correlated twin beams could be produced [194]. Also, logic gate operation [195] and many other quantum devices can be achieved with cavity-QED systems involving multilevel EIT systems [196].

Appendix A

A.1 Model of a Nonlinear Oscillator for Optical Bistability

In this appendix we discuss how the nonlinear equation used in describing the dynamical hysteresis cycle of the AOB (Eq.(7.1)) and the phenomenon of stochastic resonance (Eq.(7.11) but without the noise term) can be derived using a simple model of spring-mass system undergoing anharmonic oscillations. We then derive the steady-state solution of nonlinear equation which exhibits bistability. Nonlinear effects have been modeled by anharmonic oscillator models in several contexts. Such models have been very useful to explain some physical phenomena that include but not limited to ionospheric plasmas, vibration spectroscopy of molecules, lattice thermal expansion and lattice thermal conductivity in solids [197]. A simple mechanical system, e.g., the simple pendulum with a finite swing amplitude or angle, exhibits anharmonic behavior. Similar anharmonic behavior in another simple mechanical system consisting of a mass attached to two springs with a finite displacement has also been observed (see Fig.A.1). In the following we concentrate on this spring-mass system and demonstrate how it produces bistability under the steady-state condition. Suppose both springs are in their unstretched positions and the mass is in its equilibrium position so that there is no potential energy in the two springs. Also, gravitational energy for the mass is very small and can be neglected here. Stretching the mass from the equilibrium position causes exertion of a restoring force by the two springs [197,198]. Assuming identical springs each with a spring constant k, the net restoring force in horizontal direction is given by Hooke's law as [198]

$$F = -2k(r - t)\sin\theta, \qquad (A.1)$$

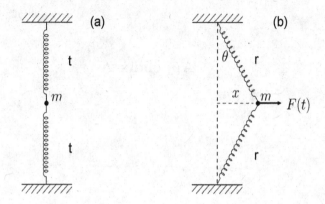

Fig. A.1 Double spring and mass system at (a) equilibrium position and (b) stretched position, respectively.

where

$$r = \sqrt{t^2 + x^2}, \quad \sin\theta = \frac{x}{r} = \frac{x}{\sqrt{t^2 + x^2}}. \tag{A.2}$$

The equilibrium length of the spring is t and the displacement of the mass from its equilibrium position is x. Thus we can write the restoring force F as [198]

$$F = -\frac{2kx}{\sqrt{t^2 + x^2}}(\sqrt{t^2 + x^2} - t) = -2kx(1 - \frac{t}{\sqrt{t^2 + x^2}}). \tag{A.3}$$

Further assuming x/t to be a small quantity and using binomial expansion we get

$$F \cong -kt\left(\frac{x}{t}\right)^3\left[1 - \frac{3}{4}\left(\frac{x}{t}\right)^2 + ...\right]. \tag{A.4}$$

By retaining only the leading term in the above expression, the quantity F becomes

$$F(x) \cong -\left(\frac{k}{t^2}\right)x^3. \tag{A.5}$$

Clearly, even for a small amplitude of oscillation ($x << t$) the restoring force is proportion to x^3 and hence it is intrinsically a nonlinear system. In practical consideration it is necessary to stretch each spring by a small displacement d just to connect it with the mass in equilibrium position. So we need to replace the intrinsic length of the spring(s) by $t + d$ and redo the above analysis, which eventually gives

$$F(x) \cong -2(kd/t)x - [k(t - d)/t^3]x^3, \tag{A.6}$$

so we can recover a linear term in this expression. The spring system is considered to be hard, provided the term $-[k(t-d)/t^3)] < 0$. Hence the system exhibits a simple harmonic motion approximately for small oscillation amplitudes. If the driving force and the linear damping terms are included in this system, then the equation of motion of the displacement x for this driven system is given by

$$m\ddot{x} + \Gamma\dot{x} = -2(kd/t)x - [k(t-d)/t^3]x^3 + A_0\cos(\omega t). \tag{A.7}$$

Note that in the above equation Γ represents linear damping coefficient and $A_0\cos(\omega t)$ is the driving force. It is convenient to define some new parameters in the above equation as $\omega_0^2 = 2kd/mt$, $\beta = k(t-d)/mt^3$, $\gamma = \Gamma/m$ and $G_0 = A_0/m$, so Eq.(A.7) can be recasted into

$$\ddot{x} + \gamma\dot{x} + \omega_0^2 x + \beta x^3 = G_0\cos(\omega t), \tag{A.8}$$

which is a standard equation of motion for a driven damped anharmonic oscillator [197–199]. The control parameter in this equation is the driving force $G_0\cos(\omega t)$. Under the limiting condition of overdamping, this equation represents Eq.(7.1) as well as Eq.(7.11) (without the noise term) used in Chapter 7. Next we define $\Delta = \omega_0^2 - \omega^2$ and substitute it in Eq.(A.8) to get

$$\ddot{x} + \gamma\dot{x} + \omega^2 x + \Delta x + \beta x^3 = G_0\cos(\omega t). \tag{A.9}$$

This is a second-order differential equation can be conveniently written into two first-order differential equations

$$y = \dot{x},$$
$$\dot{y} = \ddot{x} = -\omega^2 x + g(x,y,t), \tag{A.10}$$

in which $g(x,y,t) = -\Delta x - \gamma y - \beta x^3 + G_0\cos(\omega t)$. These equations can be solved using Van der Pol transformation [199] defined by

$$x = \cos\omega t \; p - \sin\omega t \; q,$$
$$y = -\omega\sin\omega t \; p - \omega\cos\omega t \; q, \tag{A.11}$$

from which we obtain

$$\dot{p} = \cos\omega t \times 0 - (1/\omega) \; \sin\omega t \; g(x,y,t),$$
$$\dot{q} = -\sin\omega t \times 0 - (1/\omega) \; \cos\omega t \; g(x,y,t). \tag{A.12}$$

Eq.(A.12) when averaged over one cycle yields

$$<\dot{p}> = -(1/2\omega)[\Delta q + \gamma\omega p + \frac{3\beta}{4}(p^2+q^2)q],$$
$$<\dot{q}> = -(1/2\omega)[-\Delta p + G_0 + \gamma\omega q - \frac{3\beta}{4}(p^2+q^2)p]. \tag{A.13}$$

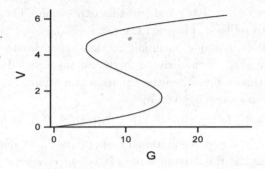

Fig. A.2 Input driving field squared amplitude G (proportional to G_0^2) vs oscillator's squared amplitude V (proportional to v^2). This curve exhibits bistable behavior in amplitude domain when $\Delta = -4.5$.

The expressions in Eq.(A.13) are coupled nonlinear differential equations and their general time-dependent solutions can be obtained numerically. However, to get bistability we are only interested in the steady-state solution of these equations. For simplicity we define p and q in terms of two new variables v and Φ, e.g., $p = 2v\cos(\Phi)$ and $q = 2v\sin(\Phi)$, and get the steady-state solutions of v from the expression [199]

$$v^2[(\Delta + 3\beta v^2)^2 + \gamma^2\omega^2] = G_0^2/4, \qquad (A.14)$$

and Φ from

$$\sin(\Phi) = -2\gamma\omega v/G_0. \qquad (A.15)$$

Equation (A.14) gives three real roots when $\Delta^2 - 3\gamma^2\omega^2 > 0$, provided $\beta \neq 0$. The bistability exhibited by the nonlinear equation (Eq.(A.8))

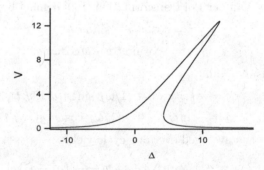

Fig. A.3 Plot of the parmeter $\Delta = \omega_0^2 - \omega^2$ vs oscillator's squared displacement amplitude V. This curve exhibits bistable behavior in frequency domain when $G = 12.5$.

under the steady-state condition can be studied by using Eq.(A.14). Such bistability behaviors in the amplitude and frequency domains are displayed in Fig.A2 and Fig.A3, respectively. Such bistability can be controlled by the parameters Δ, β, γ, and ω etc. It is possible to estimate the average power absorbed by the nonlinear oscillator during one time period as [199]

$$< P > \sim \frac{G_0^2 \gamma \omega^2}{2[(\Delta + \frac{3}{4} \frac{\beta G_0^2}{[\Delta^2 + \gamma^2 \omega^2]})^2 + \gamma^2 \omega^2]}. \qquad (A.16)$$

Clearly, the power absorbed in the nonlinear oscillator is sensitively dependent on the system parameters Δ, β, γ, ω, and G_0 etc.

Bibliography

[1] Lugiato, L. A. (1984). Theory of Optical Bistability, in E. Wolf (ed.), *Progress in Optics*, Vol. 21, (North Holland, Amsterdam), pp. 71–216.

[2] Gibbs, H. M. (1985). *Optical Bistability: Controlling Light with Light* (Academic Press, New York).

[3] Boyd, R. W. (1992). *Nonlinear Optics* (Academic Press, New York).

[4] Joshi, A. and Xiao, M. (2006). Controlling nonlinear optical processes in multi-level atomic systems, in E. Wolf (ed.), *Progress in Optics*, Vol. 49, (North Holland, Amsterdam), pp. 97–175.

[5] Joshi, A. and Xiao, M. (2010). Atomic optical bistability in two- and three-level systems: perspectives and prospects, *J. Mod. Opt.* **57**, 14/15, pp. 1196-1220.

[6] Gibbs, H. J., McCall, S. L. and Venkatesan, T. N. C. (1976). Differential gain and bistability using a sodium-filled Fabry-Perot interferometer, *Phys. Rev. Lett.* **36**, 19, pp. 1135–1138.

[7] Venkatesan, T. N. C. and McCall, S. L. (1977). Optical bistability and differential gain between 85 and 296 K in a Fabry-Perot containing ruby, *Appl. Phys. Lett.* **30**, 6, pp. 282–284.

[8] Agrarwal, G. P. and Carmichael, H. J. (1979). Optical bistability through nonlinear dispersion and absorption, *Phys. Rev. A* **19**, 5, pp. 2074–2086.

[9] Narducci, L. M., Gilmore, R., Feng, D. H., Agarwal, G. S. (1978). Quantum analysis of optical bistability and spectrum of fluctuations, *Opt. Lett.* **2**, 4, pp. 88–90.

[10] Agarwal, G. S., Narducci, L. M., Gilmore, R. and Feng, D. H. (1978). Optical bistability: A self-consistent analysis of fluctuations and spectrum of scattered light, *Phys. Rev. A* **18**, 2, pp. 620–634.

[11] Narducci, L. M., Gilmore, R., Feng, D. H. and Agarwal, G. S. (1978). Absorption spectrum of optically bistable systems, *Phys. Rev. A* **20**, 2, pp. 545–549.

[12] Lugiato, L. A., Ferina, J. D. and Narducci, L. M., (1980). Quantum-statistical treatment of the transient in absorptive optical bistability–local relaxation, *Phys. Rev. A* **22**, 1, pp. 253-260.

[13] Lugiato, L. A., Benza, V., Narducci, L. M. and Ferina, J. D. (1981). Optical bistability, instabilities and higher order bifurcations, *Opt. Commun.* **39**, 6, pp. 405–410.

[14] Lugiato, L. A., Benza, V., Narducci, L. M. and Ferina, J. D. (1983). Dressed mode description of optical bistability III. transient behaviour and higher order bifurcations, *Zeitschrift für Phys. B* **49**, 4, pp. 351-361.

[15] Orozco, L. A., Rosenberger, A. T. and Kimble, H. J. (1984). Intrinsic dynamical instability in optical bistability with two-level atoms, *Phys. Rev. Lett.* **53**, 27, pp. 2547-2550.

[16] Orozco, L. A., Kimble, H. J., Rosenberger, A. T., Lugiato, L. A., Asquini, M. L., Brambilla, M. and Narducci, L. M. (1989). Single-mode instability in optical bistability, *Phys. Rev. A* **39**, 3, pp. 1235-1252.

[17] Lambrecht, A., Giacobino, E. and Courty, J. M. (1995). Optical nonlinear dynamics with cold atoms in a cavity, *Opt. Commun.* **115**, 1/2, pp. 199–206.

[18] Arimondo, E. (1996). Coherent Population Trapping in Laser Spectroscopy, in E. Wolf (ed.), *Progress in Optics*, Vol. 35, (North Holland, Amsterdam), pp. 257–354.

[19] Harris, S. E. (1997). Electromagnetically induced transparency, *Physics Today*, **50**, 7, pp. 36-42.

[20] Marangos, J. P. (1998). Electromagnetically induced transparency, *J. Mod. Opt.*, **45**, 3, pp. 471-503.

[21] Boller, K. J., Imamoglu, A. and Harris, S. E. (1991). Observation of electromagnetically induced transparency, *Phys. Rev. Lett.* **66**, 20, pp. 2593-2596.

[22] Field, J. E., Hahn, K. H. and Harris, S. E. (1991). Observation of electromagnetically induced transparency in collisionally broadened lead vapor, *Phys. Rev. Lett.* **67**, 22, pp. 3062-3065.

[23] Li, Y. and Xiao, M. (1995). Electromagnetically induced transparency in a three-level Λ-type system in rubidium atoms, *Phys. Rev. A* **51**, 4, pp. R2703-R2706.

[24] Gea-Banacloche, J., Li, Y., Jin, S. and Xiao, M. (1995). Electromagnetically induced transparency in ladder-type inhomogeneously broadened media: Theory and experiment, *Phys. Rev. A* **51**, 1, pp. 576-584.

[25] Xiao, M., Li, Y., Jin, S. and Gea-Banacloche, J. (1995). Measurement of dispersive properties of electromagnetically induced transparency in rubidium atoms, *Phys. Rev. Lett.* **74**, 5, pp. 666-669.

[26] Hakuta, K., Marmet, L. and Stoicheff, B. P. (1991). Electric-field-induced second-harmonic generation with reduced absorption in atomic hydrogen, *Phys. Rev. Lett.* **66**, 5, pp. 596-599.

[27] Merriam, A. J., Sharpe, S. J., Shverdin, M., Manuszak, D., Yin, G. Y. and Harris, S. E. (2000). Efficient nonlinear frequency conversion in an all-resonant double-Λ system, *Phys. Rev. Lett.* **84**, 23, pp. 5308-5311.

[28] Jain, M., Xia, H., Yin, G. Y., Merriam, A. J. and Harris, S. E. (1996). Efficient nonlinear frequency conversion with maximal atomic coherence, *Phys. Rev. Lett.* **77**, 21, pp. 4326-4329.

[29] Yan, M. and Rickey, E. G. and Zhu, Y. (2001). Nonlinear absorption by quantum interference in cold atoms, *Opt. Lett.* **26**, 8, pp. 548-550.

[30] Hemmer, P. R., Katz, D. P., Donoghue, J., Cronin-Golomb, M., Shahriar, M. S. and Kumar, P. (1995). Efficient low-intensity optical phase conjugation based on coherent population trapping in sodium, *Opt. Lett.* **20**, 9, pp. 982-984.

[31] Li, Y. and Xiao, M. (1996). Enhancement of nondegenerate four-wave mixing based on electromagnetically induced transparency in rubidium atoms, *Opt. Lett.* **21**, 14, pp. 1064-1066.

[32] Lu, B., Burkett, W. H. and Xiao, M. (1998). Nondegenerate four-wave mixing in a double-Λ system under the influence of coherent population trapping, *Opt. Lett.* **23**, 10, pp. 804-806.

[33] Wang, H., Goorskey, D. J. and Xiao, M. (2001). Enhanced Kerr nonlinearity via atomic coherence in a three-level atomic system, *Phys. Rev. Lett.* **87**, 7, 073601.

[34] Wang, H., Goorskey, D. and Xiao, M. (2001). Dependence of enhanced Kerr nonlinearity on coupling power in a three-level atomic system, *Opt. Lett.* **27**, 4, pp. 258-260.

[35] Mlynek, J., Mitschke, F., Deserno, R. and Lange, W. (1984). Optical bistability from three-level atoms with the use of a coherent nonlinear mechanism, *Phys. Rev. A* **29**, 3, pp. 1297-1303.

[36] Harshawardhan, H. and Agarwal, G. S. (1996). Controlling optical bistability using electromagnetic-field-induced transparency and quantum interferences, *Phys. Rev. A* **53**, 3, pp. 1812-1817.

[37] Gong, S. -Q., Du, S. -D. and Xu, Z. -Z. (1997). Optical bistability via atomic coherence, *Phys. Lett. A* **226**, 5, pp. 293-297.

[38] Walls, D. F. and Zoller, P. (1980). A coherent nonlinear mechanism for optical bistability from three level atoms, *Opt. Commun.* **34**, 2, pp. 260-264.

[39] Savage, C. M., Carmichael, H. J. and Walls, D. F. (1982). Optical multistability and self oscillations in three level systems, *Opt. Commun.* **42**, 3, pp. 211-216.

[40] Joshi, A., Yang, W. and Xiao, M. (2003). Effect of quantum interference on optical bistability in the three-level V-type atomic system, *Phys. Rev. A* **68**, 1, 015806.

[41] Joshi, A., Yang, W. and Xiao, M. (2003). Effect of spontaneously generated coherence on optical bistability in three-level Λ-type atomic system, *Phys. Lett. A* **315**, 3/4, pp. 203-207.

[42] Joshi, A., Brown, A., Wang, H. and Xiao, M. (2003). Controlling optical bistability in a three-level atomic system, *Phys. Rev. A* **67**, 4, 041801(R).

[43] Joshi, A. and Xiao, M. (2003). Optical multistability in three-level atoms inside an optical ring cavity, *Phys. Rev. Lett.* **91**, 14, 143904.

[44] Joshi, A., Yang, W. and Xiao, M. (2004). Hysteresis loop with controllable shape and direction in an optical ring cavity, *Phys. Rev. A* **70**, 4, 041802(R).

[45] Joshi, A., Yang, W. and Xiao, M. (2004). Dynamical hysteresis in a three-level atomic system, *Opt. Lett.* **30**, 8, pp. 905-907.

[46] Yang, W., Joshi, A. and Xiao, M. (2004). Controlling dynamic instability of three-level atoms inside an optical ring cavity, *Phys. Rev. A* **70**, 3, 033807.

[47] Yang, W., Joshi, A. and Xiao, M. (2005). Chaos in an electromagnetically induced transparent medium inside an optical cavity, *Phys. Rev. Lett.* **95**, 9, 093902.

[48] Joshi, A. and Xiao, M. (2006). Stochastic resonance in atomic optical bistability, *Phys. Rev. A* **74**, 1, 013817.

[49] Joshi, A. and Xiao, M. (2008). Noise-induced switching via fluctuating atomic coherence in an optical three-level bistable system, *J. Opt. Soc. Am. B* **25**, 12, pp. 2015-2020.

[50] Allen, L. and Eberly, J. H. (1975). *Optical Resonance and Two-level Atoms* (Wiley, New York).

[51] Cohen-Tannoudji, C., Dupont-Roc, J. and Grynberg, G. (1989). *Photons and Atoms: Introduction to Quantum Electrodynamics* (Wiley, New York).

[52] Cohen-Tannoudji, C., Dupont-Roc, J. and Grynberg, G. (1992). *Atom-Photon Interactions: Basic Processes and Applications* (Wiley, New York).

[53] Shore, B. W. (1990). *The Theory of Coherent Atomic Excitation*, (Wiley, New York).

[54] Demtröder, W. (1981). *Laser Spectroscopy: Basic Concepts and Instrumentation* (Springer-Verlag, Berlin and Heidelberg).

[55] Preston, D. W. (1996). Doppler-free saturated absorption: Laser spectroscopy, *Am. J. Phys.* **64**, 11, pp. 1432–1436.

[56] *Doppler-Free Saturated Absorption Spectroscopy: Laser Spectroscopy,* http://optics.colorado.edu/~kelvin/classes/opticslab/ LaserSpectroscpy6.doc.pdf

[57] Preston, D. W. and Wieman, C. E. *Doppler-Free Saturated Absorption Spectroscopy: Laser Spectroscopy,* http://socrates.berkeley.edu/ ~phylabs/adv/ReprintsPDF/MNO%20Reprints/04%20-%20Doppler% 20Free%20Saturated%20Absoprtion.pdf

[58] Born, M. and Wolf, E. (1975). *Principles of Optics* (Pergamon Press, Oxford).

[59] http://en.wikipedia.org/wiki/Optical$_$cavity

[60] Yariv, A. (1989). *Quantum Electronics*, 3rd edn. (John Wiley and Sons, NJ).

[61] Betzler, K. http://www.fen.bilkent.edu.tr/%7Eaykutlu/msn513/ fibersensors/fabryperot.pdf

[62] Goorskey, D. J., Wang, H., Burkett, W. H. and Xiao, M. (2002). Effects of a highly dispersive atomic medium inside an optical ring cavity, *J. Mod. Opt.* **49**, 1/2, 305-317.

[63] Fano, U. (1961). Effects of configuration interaction on intensities and phase shifts, *Phys. Rev.* **124**, 6, pp. 1866–1878.

[64] Gray, H. R., Whitley, R. M. and Stroud, Jr., C. R. (1978). Coherent trapping of atomic populations, *Opt. Lett.* **3**, 6, pp. 218–220.

[65] Balian, R., Haroche, S. and Liberman, S. (eds.) (1977). *Frontiers in Laser spectroscopy*, Vol.1, (North Holland, Amsterdam).

[66] Harris S. E. (1989). Lasers without inversion: Interference of lifetime-broadened resonances, *Phys. Rev. Lett.* **62**, 9, pp. 1033–1036.

[67] Harris, S. E., Field, J. E. and Imamoglu, A. (1990). Nonlinear optical processes using electromagnetically induced transparency, *Phys. Rev. Lett.* **64**, 10, pp. 1107-1110.

[68] Li, Y. Q. and Xiao, M. (1995). Observation of quantum interference between dressed states in an electromagnetically induced transparency, *Phys. Rev. A* **51**, 6, pp. 4959-4962.

[69] Li, Y. Q. and Xiao, M. (1995). Transient properties of an electromagnetically induced transparency in three-level atoms, *Opt. Lett.* **20**, 13, pp. 1489-1491.

[70] Harris, S. E. (1993). Electromagnetically induced transparency with matched pulses, *Phys. Rev. Lett.* **70**, 5, pp. 552-555.

[71] Harris, S. E. and Luo, Z. (1995). Preparation energy for electromagnetically induced transparency, *Phys. Rev. A* **52**, 2, pp. R928-R931.

[72] Li, Y., Jin, S. and Xiao, M. (1995). Observation of an electromagnetically induced change of absorption in multilevel rubidium atoms, *Phys. Rev. A* **51**, 3, pp. R1754-R1757.

[73] Kash, M. M., Sautenkov, V. A., Zibrov, A. S., Hollberg, L., Welch, G. R., Lukin, M. D., Rostovtsev, Y., Fry, E. S. and Scully, M. O. (1999). Ultraslow group velocity and enhanced nonlinear optical effects in a coherently driven hot atomic gas, *Phys. Rev. Lett.* **82**, 26, pp. 5229-5232.

[74] Hau, L. V., Harris, S. E., Dutton, Z. and Behroozi, C. H. (1999). Light speed reduction to 17 metres per second in an ultracold atomic gas, *Nature* **397**, 6720, pp. 594–598.

[75] Yan, M., Rickey, E. G. and Zhu, Y. (2001). Observation of doubly dressed states in cold atoms, *Phys. Rev. A* **64**, 1, 013412.

[76] Wang, H., Goorskey, D. J. and Xiao, M. (2002). Atomic coherence induced Kerr nonlinearity enhancement in Rb vapour, *J. Mod. Opt.* **49**, 3/4, pp. 335-347.

[77] Budker, D., Kimball, D. F., Rochester, S. M. and Yashchuk, V. V. (1999). Nonlinear magneto-optics and reduced group velocity of Light in atomic vapor with slow ground state relaxation, *Phys. Rev. Lett.* **83**, 9, pp. 1767–1770.

[78] Moseley, R. R., Shepherd, S., Fulton, D. J., Sinclair, B. D. and Dunn, M. H. (1995). Spatial consequences of electromagnetically induced transparency: Observation of electromagnetically induced focusing, *Phys. Rev. Lett.* **74**, 5, pp. 670–673.

[79] Chang, H., Du, Y., Yao, J., Xie, C. and Wang, H. (2004). Observation of cross-phase shift in hot atoms with quantum coherence, *Europhys. Lett.* **65**, 4, 485-490.

[80] Bonifacio, R. and Lugiato, L. A. (1976). Cooperative effects and bistability for resonance fluorescence, *Opt. Commun.* **19**, 2, pp. 172–176.

[81] Gripp, J. and Orozco, L. A. (1996). Evolution of the vacuum Rabi peaks in a many-atom system, *Quant. Semiclass. Opt.* **8**, 4, pp. 823–836.

[82] Gripp, J., Mielke, S. L., Orozco, L. A. and Carmichael, H. J. (1997). Anharmonicity of the vacuum Rabi peaks in a many-atom system, *Phys. Rev. A* **54**, 5, pp. R3746–R3749.

[83] Gripp, J., Mielke, S. L. and Orozco, L. A. (1997). Evolution of the vacuum Rabi peaks in a detuned atom-cavity system, *Phys. Rev. A* **56**, 4, pp. 3262–3273.

[84] Zhu, Y., Gauthier, D. J., Morin, S. E., Wu, Q., Carmichael, H. J. and Mossberg, T. W. (1990). Vacuum Rabi splitting as a feature of linear-dispersion theory: Analysis and experimental observations, *Phys. Rev. Lett.* **64**, 21, pp. 2499–2502.

[85] Szöke, A., Daneu, V., Goldhar, J. and Kurnit, N. A. (1969). Bistable optical element and its applications, *Appl. Phys. Lett.* **15**, 11, pp. 376–378.

[86] Bonifacio, R. and Lugiato, L. A. (1978). Mean field model for absorptive and dispersive bistability with inhomogeneous broadening, *Lettere Al Nuovo Cimento* **21**, 15, pp. 517–521.

[87] Rosenberger, A. T., Orozco, L. A. and Kimble, H. J. (1983). Observation of absorptive bistability with two-level atoms in a ring cavity, *Phys. Rev. A* **28**, 4, pp. 2569–2572.

[88] Orozco, L. A., Rosenberger, A. T. and Kimble, H. J. (1987). Optical bistability in the mixed absorptive-dispersive regime with two-state atoms, *Phys. Rev. A* **36**, 7, pp. 3248–3252.

[89] Scully, M. O. and Zubairy, M. S. (1997). *Quantum Optics* (Cambridge University Press, Cambridge).

[90] Zibrov, A. S., Lukin, M. D., Hollberg, L., Nikonov, D. E., Scully, M. O., Robinson, H. G. and Velichansky, V. L. (1996). Experimental demonstration of enhanced index of refraction via quantum interference in Rb, *Phys. Rev. Lett.* **76**, 21, pp. 3935–3998.

[91] Truscott, A. G., Friese, M. E. J., Heckenberg, N. R. and Rubinsztein-Dunlop, H. (1999). Optically written waveguide in an atomic vapor, *Phys. Rev. Lett.* **82**, 7, pp. 1438–1441.

[92] Liu, C., Dutton, Z., Behroozi, C. H. and Hau, L. V. (2001). Observation of coherent optical information storage in an atomic medium using halted light pulses, *Nature*, **409**, 6819, pp. 490–493.

[93] Zhou, P. and Swain, S. (1998). Dynamical suppression of the Autler-Townes doublet in the presence of a cavity, *Phys. Rev. A*, **57**, 5, pp. 3781-3787.

[94] Lukin, M. D., Fleischhauer, M., Scully, M. O. and Velichansky, V. L. (1998). Intracavity electromagnetically induced transparency, *Opt. Lett.* **23**, 4, pp. 295–297.

[95] Wang, H., Goorskey, D. J., Burkett, W. H. and Xiao, M. (2000). Cavity-linewidth narrowing by means of electromagnetically induced transparency, *Opt. Lett.* **25**, 23, pp. 1732–1734.

[96] Harris, S. E. and Hau, L. U. (1999). Nonlinear optics at low light levels, *Phys. Rev. Lett.* **82**, 23, pp. 4611–4614.

[97] Imamoglu, A., Schmidt, H., Woods, G. and Deutsch, M. (1997). Strongly interacting photons in a nonlinear Cavity, *Phys. Rev. Lett.* **79**, 8, pp. 1467–1470.

[98] Jin, S. Z., Li, Y. Q. and Xiao, M. (1996). Single mode diode laser with a large frequency scanning range based on weak grating feedback, *Appl. Opt.* **35**, 9, pp. 1436–1441.

[99] Wang, L. J., Kuzmich, A. and Dogariu, A. (2000). Gain-assisted superluminal light propagation, *Nature* **406**, 6793, pp. 277–279.

[100] Akushin, A. M., Barreiro, S. and Lezema, A. (1999). Steep anomalous dispersion in coherently prepared Rb Vapor, *Phys. Rev. Lett.* **83**, 21, pp. 4277–4280.

[101] Pati, G. S., Salit, M., Salit, K. and Shahriar, M. S. (2007). Demonstration of a tunable-bandwidth white-light interferometer using anomalous dispersion in atomic vapor, *Phys. Rev. Lett.* **99**, 13, 133601.

[102] Wu, H. and Xiao, M. (2007). Cavity linewidth narrowing and broadening due to competing linear and nonlinear dispersions, *Opt. Lett.* **32**, 21, pp. 3122–3124.

[103] Wu, H. and Xiao, M. (2008). White-light cavity with competing linear and nonlinear dispersions, *Phys. Rev. A* **77**, 3, 031801.

[104] O'Keefe, A., Scherer, J. J., Paul, J. B. and Saykally, R. J., Cavity Ringdown Laser Spectroscopy (CRDS): History, Development, and Applications, http://www.Igrinc.com/publications/acs.pdf

[105] Anderson, D. Z., Frisch, J. C. and Masser, C. S. (1984). Mirror reflectometer based on optical cavity decay time, *Appl. Opt.* **23**, 8, pp. 1238–1245.

[106] Vallance, C. (2005). Innovations in cavity ringdown spectroscopy, *New J. Chem.* **29**, 7, pp. 867-874.

[107] O'Keefe, A. and Deacon, D. A. G. (1988). Cavity ring-down optical spectrometer for absorption measurements using pulsed laser sources, *Rev. Sci. Instrum.* **59**, 12, pp. 2544–2551.

[108] Yang, W., Joshi, A. and Xiao, M. (2004). Enhancement of the cavity ringdown effect based on electromagnetically induced transparency, *Opt. Lett.* **29**, 18, pp. 2133-2135.

[109] Zalicki, P. and Zare, R. N. (1995). Cavity ring-down spectroscopy for qunatitative absorption measurements, *J. Chem. Phys.* **102**, 7, pp. 2708–2717.

[110] Yu, T. and Lin, M. C. (1993). Kinetics of phenyl radical reactions studied by the cavity-ring-down method, *J. Am. Chem. Soc.* **115**, 10, pp. 4371–4372.

[111] Romanini, D. and Lehmann, K. K. (1995). Cavity ring-down overtone spectroscopy of HCN, $H^{13}CN$ and $HC^{15}N$, *J. Chem. Phys.* **102**, 2, pp. 633–642.

[112] Meijer, G., Boogaarts, M. G. H., Jongma, R. T., Parker, D. H. and Wodtke, A. M. (1994). Coherent cavity ring down spectroscopy, *Chem. Phys. Lett.* **217**, 1/2, pp. 112–116.

[113] Bakowski, B., Corner, L., Hancock, G., Kotchie, R., Peverall, R. and Ritchie, G. A. D. (2002). Cavity-enhanced absorption spectroscopy with a rapidly swept diode laser, *Appl. Phys. B* **75**, 6/7, pp. 745–750.

[114] Berden, G., Peeters, R. and Meijer, G. (2000). Cavity ring-down spectroscopy: Experimental schemes and applications, *Int. Rev. Phys. Chem.* **19**, 4, pp. 565–607.

[115] An, K., Yang, C., Dasari, R. R. and Feld, M. S. (1995). Cavity ring-down technique and its application to the measurement of ultraslow velocities, *Opt. Lett.* **20**, 9, pp. 1068–1070.

[116] Poirson, J., Bretenaker, F., Vallet, M. and Floch, A. L. (1997). Analytical and experimental study of ringing effects in a Fabry-Perot cavity: Application to the measurement of high finesses, *J. Opt. Soc. Am. B* **14**, 11, pp. 2811–2817.

[117] Chang, H., Wu, H., Xie, C. and Wang, H. (2004). Controlled shift of optical bistability hysteresis curve and storage of optical signals in a four-level atomic system, *Phys. Rev. Lett.* **93**, 21, 213901.

[118] Vadhawan, V. K. (2002), *Introduction to Ferroic materials* (Gordan and Breach, UK).

[119] Jewell, J. L., Gibbs, H. M., Tarng, S. S., Gossard, A. C. and Wiegmann, W. (1982). Regenerative pulsations from an intrinsic bistable optical device, *Appl. Phys. Lett.* **40**, 4, pp. 291–293.

[120] Dagenais, M. and Sharfin, W. F. (1985). Linear- and nonlinear-optical properties of free and bound excitons in CdS and applications in bistable devices, *J. Opt. Soc. Am. B* **2**, 7, pp. 1179–1187.

[121] Giusfredi, G., Salieri, P., Cecchi, S. and Arecchi, F. T. (1985). Multistability in a sodium filled Fabry-Perot: D1 line, hyperfine and Zeeman pumping, *Opt. Commun.* **54**, 1, pp. 39–46.

[122] Giacobino, E. (1985). Tristability and bifurcations in sodium vapor, *Opt. Commun.* **56**, 4, pp. 249–254.

[123] Hamilton, M. W., Sandle, W. J., Chilwell, J. T., Satchell, J. S. and Warrington, D. M. (1983). Observation of polarization switching: D1 line of sodium in a Fabry-Perot, *Opt. Commun.* **48**, 3, pp. 190–194.

[124] Mitschke, F., Deserno, R., Lange, W. and Mlynek, J. (1986). Magnetically induced optical self-pulsing in a nonlinear resonator, *Phys. Rev. A* **33**, 5, pp. 3219–3231.

[125] Kitano, M., Yabuzaki, T. and Ogawa, T. (1981). Optical tristability, *Phys. Rev. Lett.* **46**, 14, pp. 926–929.

[126] Cecchi, S., Giusfredi, G., Petriella, E. and Salieri, P. (1982). Observation of optical tristability in sodium vapors, *Phys. Rev. Lett.* **49**, 26, pp. 1928–1931.

[127] Arecchi, F. T., Kurmann, J. and Politi, A. (1983). A new class of optical multistabilities and instabilities induced by atomic coherence, *Opt. Commun.* **44**, 6, pp. 421–425.

[128] Mair, A., Hager, J., Phillips, D. F., Walsworth, R. L. and Lukin, M. D. (2002). Phase coherence and control of stored photonic information, *Phys. Rev. A* **65**, 3, 031802(R).

[129] http://classes.entom.wsu.edu/529/stability.htm

[130] Lugiato, L. A., Narducci, L. M., Bandy, D. K. and Pennise, C. A. (1982). Self-pulsing and chaos in a mean-field model of optical bistability, *Opt. Commun.* **43**, 4, pp. 281–286.

[131] Haken, H. (1978). *Synergetics* (Springer-Verlag, Berlin).

[132] Haken, H. (1975). Generalized Ginzburg-Landau equations for phase transition-like phenomena in lasers, nonlinear optics, hydrodynamics and chemical reactions, *Zeitschrift für Phys. B* **21**, 1, pp. 105–114.

[133] Ikeda, I. (1979). Multiple-valued stationary states and its instability of the transmitted light by a ring cavity system, *Opt. Commun.* **30**, 2, pp. 257–261.

[134] Segard, B., Macke, B., Lugiato, L. A., Prati, F. and Brambilla, M. (1989). Multimode instability in optical bistability, *Phys. Rev. A* **39**, 2, pp. 703–722.

[135] Lugiato, L. A., Horowicz, R. J., Strini, G. and Narducci, L. M. (1984). Instabilities in passive and active optical systems with a Gaussian transverse intensity profile, *Phys. Rev. A* **30**, 3, pp. 1366–1376.

[136] Carmichael, H. J., Snapp, R. R. and Schieve, W. C. (1982). Oscillatory instabilities leading to optical turbulence in a bistable ring cavity, *Phys. Rev. A* **26**, 6, pp. 3408–3422.

[137] Asquini, M. L., Lugiato, L. A., Carmichael, H. J. and Narducci, L. M. (1986). Off-resonant-mode instabilities in mixed absorptive and dispersive optical bistabilites, *Phys. Rev. A* **33**, 1, pp. 360–374.

[138] Firth, W. J., Harrison, R. G. and Al-Saidi, I. A. (1986). Instabilities and routes to chaos in passive all-optical resonators containing a molecular gas, *Phys. Rev. A* **33**, 4, pp. 2449–2460.

[139] Bowden, C. M., Ciftan, M. and Robl, H. (eds.) (1981). *Optical bistability* (Plenum Press, NY).

[140] Lugiato, L. A. (1981). Many-mode quantum statistical theory of optical bistability, *Zeitschrift für Phys. B* **41**, 1, pp. 85–94.

[141] Ikeda, K. and Akimoto, O. (1982). Instability leading to periodic and chaotic self-pulsations in a bistable optical cavity, *Phys. Rev. Lett.* **48**, 9, pp. 617–620.

[142] Wang, H., Goorskey, D. J. and Xiao, M. (2001). Bistability and instability of three-levl atoms inside an optical cavity, *Phys. Rev. A* **65**, 1, 011801(R).

[143] Parker, T. S. and Chua, L. O. (1987). Chaos : A Tutorial for engineers, *Proceedings of the IEEE* **75**, 8, pp. 982–1008.

[144] Van Wiggeren, G. D. and Roy, R. (1998). Optical communication with chaotic waveforms, *Phys. Rev. Lett.* **81**, 16, pp. 3547–3550.

[145] Hall, G. M. and Gauthier, D. J. (2002). Experimental control of cardiac muscle alternans, *Phys. Rev. Lett.* **88**, 19, 198102.

[146] Arrechi, F. T. and Harrison, R. G. (ed.) (1987). *Instabilities and Chaos in Quantum Optics*, Springer Series in Synergetics Vol. 34 (Springer, Berlin).

[147] Moller, M. and Lange, W. (1994). Radiation trapping : An alternative mechnism for chaos in a nonlinear optical resonantor, *Phys. Rev. A* **49**, 5, pp. 4161-4169.

[148] Midavaine, T., Dangoisse, D. and Glorieux, P. (1985). Observation of chaos in a frequncy-modulated CO_2 laser, *Phys. Rev. Lett.* **55**, 19, pp. 1989-1992.

[149] Moloney, J. V., Uppal, J. S. and Harrison, R. G. (1987). Orgin of chaotic relaxtion oscillations in an optically pumped molecular laser, *Phys. Rev. Lett.* **59**, 25, pp. 2868-2871.

[150] Gibbs, H. M., Hopf, F. A., Kaplan, D. L. and Shoemaker, R. L. (1981). Observation of chaos in optical bistability, *Phys. Rev. Lett.* **46**, 7, pp. 474–477.

[151] Roy, R., Murphy, Jr., T. W., Maier, T. D., Gills, Z. and Hunt, E. R. (1992). Dynamical control of a chaotic laser : Experimental stabilization of a globally coupled system, *Phys. Rev. Lett.* **68**, 9, pp. 1259-1262.

[152] Milonni, P. W., Shih, M. -L. and Ackerhalt, J. R. (1987). *Chaos in laser-matter interactions*, World Scientific Lecture Notes in Physics, Vol. 6 (World Scientific, Singapore).

[153] Uppal, J. S., Harrison, R. G. and Moloney, J. V. (1987). Gain, dispersion, and emission characteristics of three-level molecular laser amplifier and oscillator systems, *Phys. Rev. A* **36**, 10, pp. 4823-4834.

[154] Sandri, M. (1996). Numerical calculation of Lyapunov cxponents, *The Mathematica Journal* **6**, 3, pp. 78–84.

[155] Papadimitriou, G. I. (2003). Optical switching : Switch fabrics, techniques, and architectures, *J. Lightwave Technol.* **21**, 2, pp. 384–405.

[156] http://www.topdocumentinfoblog.info/communications /understanding-the-basics-of-all-optical-switching

[157] http://www.phy.duke.edu/research /photon/qelectron/proj/switch/intro.php

[158] Jensen, S. (1982). The nonlinear coherent coupler, *IEEE J. Quant. Electron.* **18**, 10, pp. 1580–1583.

[159] Wang, H., Goorskey, D. and Xiao, M. (2002). Controlling light by light with three-level atoms inside an optical cavity, *Opt. Lett.* **27**, 15, pp. 1354–1356.

[160] Wang, H., Goorskey, D. and Xiao, M. (2002). Controlling the cavity field with enhanced Kerr nonlinearity in three-level atoms, *Phys. Rev. A* **65**, 5, 051802(R).

[161] Brown, A., Joshi, A. and Xiao, M. (2003). Controlled steady-state switching in optical bistability, *Appl. Phys. Lett.* **83**, 7, pp. 1301–1303.

[162] Yan, M., Rickey, E. G. and Zhu, Y. (2001). Observation of absorptive photon switching by quantum interference, *Phys. Rev. A* **64**, 4, 041801.

[163] Harris, S. E., and Yamamoto, Y. (1998). Photon switching by Quantum interference, *Phys. Rev. Lett.* **81**, 17, pp. 3611–3614.

[164] Jung, P., Gray, G., Roy, R. and Mandel, P. (1990). Scaling law for dynamical hysteresis, *Phys. Rev. Lett.* **65**, 15, pp. 1873–1876.

[165] Goldsztein, G. H., Broner, F. and Strogatz, S. H. (1997). Dynamical hysteresis without static hysteresis : scaling laws and asymptotic expansions, *SIAM J. Appl. Math.* **57**, 4, pp. 1163–1187.

[166] Hohl, A., van der Linden, H. J. C., Roy, R., Goldsztein, G., Broner, F. and Strogatz, S. H. (1995). Scaling laws for dynamical hysresis in a multi-dimensional laser system, *Phys. Rev. Lett.* **74**, 12, pp. 2220–2223.

[167] Yamada, M. (1986). Theory of mode competition noise in semiconductor injection lasers, *IEEE J. Quant. Electron.* **22**, 7, pp. 1052–1059.

[168] Novikova, I., Zibrov, A. S., Phillips, D. F., Andre, A. and Walsworth, R. L. (2004). Dynamic optical bistability in resonantly enhanced Raman generation, *Phys. Rev. A* **69**, 6, 061802(R).

[169] Risken, H., Savage, C., Haake, F. and Walls, D. F. (1987). Quantum tunneling in dispersive optical bistability, *Phys. Rev. A* **35**, 4, pp. 1729–1739.

[170] Thorwart, M. and Jung, P. (1997). Dynamical hysteresis in bistable quantum systems, *Phys. Rev. Lett.* **78**, 13, pp. 2503–2506.

[171] Turukhin, A. V., Sudarshanam, V. S., Shahriar, M. S., Musser, J. A., Ham, B. S. and Hemmer, P. R. (2002). Observation of ultraslow and stored light pulses in a solid, *Phys. Rev. Lett.* **88**, 2, 023602.

[172] Phillips, M., Wang, H., Rumyantsev, I., Kwong, N. H., Takayama, R. and Binder, R. (2003). Electromagnetically induced transparency in semiconductors via biexciton coherence, *Phys. Rev. Lett.* **91**, 18, 183602.

[173] Gammaitoni, L., Hanggi, P., Jung, P. and Marchesoni, F. (1998). Stochastic resonance, *Rev. Mod. Phys.* **70**, 1, pp. 223–287.

[174] Wellens, T., Shatokhin, V. and Buchleitner, A. (2004). Stochastic resonance, *Rep. Prog. Phys.* **67**, 1, pp. 45–105.

[175] Fauve, S. and Heslot, F. (1983). Stochastic resonance in a bistable system, *Phys. Lett. A* **97**, 1/2, pp. 5–7.

[176] McNamara, B., Wiesenfeld, K. and Roy, R. (1988). Observation of stochastic resonance in a ring laser, *Phys. Rev. Lett.* **60**, 25, pp. 2626–2629.

[177] Grohs, J., Apanasevich, S., Jung, P., Issler, H., Burak, D. and Klingshirn, C. (1994). Noise-induced switching and stochastic resonance in optically nonlinear CdS crystals, *Phys. Rev. A* **49**, 3, pp. 2199–2202.

[178] Wiesenfeld, K. and Moss, F. (1995). Stochastic resonance and the benfits of noise: from ice ages to crayfish and SQUIDs, *Nature* **373**, 6509, pp. 33–36.

[179] Hanggi, P., Talkner, P. and Borkovec, M. (1990). Reaction-rate theory: fifty years after Kramers, *Rev. Mod. Phys.* **62**, 2, pp. 251–341.

[180] Kramers, H. (1940). Brownian motion in a field of force and the diffusion model of chemical reactions, *Physica (Utrecht)* **7**, 4, pp. 284–304.

[181] Gammaitoni, L., Menichella-Saetta, E., Santucci, S., Marchesoni, F. and Presilla, C. (1989). Periodically time-modulated bistable systems : stochastic resonance, *Phys. Rev. A* **40**, 4, pp. 2114–2119.

[182] Wu, H., Joshi, A., and Xiao, M. (2007). Stochastic resonance with multiplicative noise in a three-level atomic bistable system, *J. Mod. Opt.* **54**, 16/18, pp. 2441–2450.

[183] Vemuri, G. and Roy, R. (1989). Stochastic resonance in a bistable ring laser, *Phys. Rev. A* **39**, 9, pp. 4668-4674.

[184] McCall, S., Ovadia, S., Gibbs, H., Hopf, F. and Kaplan, D. (1985). Statistical fluctuations in optical bistability induced by shot noise, *IEEE J. Quant. Electron.* **21**, 9, pp. 1441-1446.

[185] Dykman, M. I., Golubev, G. P., Luchinsky, D. G., Velikovich, A. L. and Tsuprikov, S. V. (1991). Fluctuational transitions and related phenomena in a passive all-optical bistable system, *Phys. Rev. A* **44**, 4, pp. 2439-2449.

[186] Filipowicz, P., Garrison, J. C., Meystre, P. and Wright, E. M. (1987). Noise-induced switching of photonic logic elements, *Phys. Rev. A* **35**, 3, pp. 1172–1180.

[187] Bonifacio, R. and Lugiato, L. A. (1976). Optical bistability and cooperative effects in resonance fluorescence, *Phys. Rev. A* **18**, 3, pp. 1129–1144.

[188] Orozco, L. A., Raizen, M. G., Xiao, M., Brecha, R. J. and Kimble, H. J. (1987). Squeezed state generation in optical bistability, *J. Opt. Soc. Am. B* **4**, 10, pp. 1490-1500.

[189] Benenti, G., Casati, G. and Strini, G. (2004). *Principles of Quantum computation and Information: Vol. 1 - Basic Concepts* (World Scientific, New Jersey).

[190] Duan, L. -M., Lukin, M. D., Cirac, J. I. and Zoller, P. (2001). Long-distance quntum communication with atomic ensembles and linear optics, *Nature* **414**, 6862, pp. 413–418.

[191] van der Wal, C. H., Eisaman, M. D., Andre, A., Walsworth, R. L., Phillips, D. F., Zibrov, A. S. and Lukin, M. D. (2003). Atomic memory for correlated photon states, *Science* **301**, 5630, pp. 196–200.

[192] Kuzmich, A., Bowen, W. P., Boozer, A. D., Boca, A., Chou, C. W., Duan, L. -M. and Kimble, H. J. (2003). Generation of nonclassical photon pairs for scalable quantum communications with atomic samples, *Nature* **423**, 6941, pp. 731–734.

[193] Thompson, J. K., Simon, J., Loh, H. and Vuletic, V. (2006). A high-brightness source of narrowband, identical-photon pairs, *Science*, **313**, 5783, pp. 74–77.

[194] Wu, H. and Xiao, M. (2009). Bright correlated twin beams from an atomic ensemble in the optical cavity, *Phys. Rev. A* **80**, 6, 063415.

[195] Joshi, A. and Xiao, M. (2005). Phase gate with a four-level inverted-Y system, *Phys. Rev. A* **72**, 6, 062319.

[196] Joshi, A. and Xiao, M. (2006). Three-qubit quantum-gate operation in a cavity QED system, *Phys. Rev. A* **74**, 5, 052318.

[197] Thomchick, J., McKelvey, J. P. (1978). Anharmonic vibrations of an ideal Hooke's law oscillator, *Am. J. Phys.* **46**, 1, pp. 40–45.

[198] Marion, J. B., Thornton, S. P. (1995). *Classical Dynamics of Particles and Systems* (Saunders, Fort Worth).

[199] Batista, A. A., Oliveira, F. A. and Nazareno, H. N. (2008). Duffing oscillators: Control and memory effects, *Phys. Rev. E* **77**, 6, 066216.

Index